TEACHING SCIENCE IN CULTURALLY RELEVANT WAYS

IDEAS FROM SINGAPORE TEACHERS

TEACHING SCIENCE IN CULTURALLY RELEVANT WAYS

IDEAS FROM SINGAPORE TEACHERS

Editors

Teo Tang Wee
Nanyang Technological University, Singapore

Khoh Rong Lun
Temasek Junior College, Singapore

Published by

World Scientific Publishing Co. Pte. Ltd.

5 Toh Tuck Link, Singapore 596224

USA office: 27 Warren Street, Suite 401-402, Hackensack, NJ 07601

UK office: 57 Shelton Street, Covent Garden, London WC2H 9HE

Library of Congress Cataloging-in-Publication Data

Teo, Tang Wee.

   Teaching science in culturally relevant ways : ideas from Singapore teachers / by Tang Wee Teo (Nanyang Technological University, Singapore) and Rong Lun Khoh (Temasek Junior College, Singapore).

   pages cm

   ISBN 978-9814618175

   1. Science--Study and teaching--Singapore. 2. Science--Study and teaching--Activity programs--Singapore. 3. Science teachers--Singapore--Attitudes. 4. Curriculum planning--Singapore. 5. Educational innovations--Singapore. 6. Culturally relevant pedagogy--Singapore. 7. Cultural pluralism--Singapore. I. Title.

   Q183.4.S45T46 2015

   507.1'05957--dc23

                                         2014024523

British Library Cataloguing-in-Publication Data

A catalogue record for this book is available from the British Library.

In-house Editors: Lum Pui Yee/Dipasri Sardar

Typeset by Stallion Press

Email: enquiries@stallionpress.com

Printed in Singapore

Dedicated to

Singapore Science Educators

Acknowledgements

The editors and book chapter authors would like to express our utmost gratitude to the school key personnel, teachers, and laboratory support staff from the following institutions for their support of our work:

National Institute of Education

Temasek Junior College

Assumption English School

Changkat Changi Secondary School

CHIJ Saint Nicholas Girls' School (Secondary)

Chung Cheng High School (Main)

Hwa Chong Institution

Marsiling Secondary School

Nan Hua High School

Nanyang Girls' High School

Nanyang Junior College

Pasir Ris Secondary School

Saint Andrew's Secondary School

Woodlands Secondary School

And.......

Special thanks to our family and friends for their support on our work.

Contents

Foreword

One can be forgiven for thinking that the book is on food as the first eight chapters in the book are all related to food; even the ninth chapter is indirectly related to food! However, this book focuses on scientific inquiry, produced by science educators who came together to work on a resource book which is culturally relevant to students and teachers in Singapore. Students' and teachers' interest will be piqued by the inquiry activities in the book involving familiar food such as dragon fruit, tea, coconut, durian, spices and food dyes, daily life objects such as T-shirts and disposable tableware, and common activities such as the shuttle run. The activities suggested can be carried out in school with existing reagents and equipment, and the techniques involved are within the capabilities of students. Thus, I have no doubts that doing science and learning the science behind the activities will be a thrill to students and a great learning experience.

Dr. TAN Kim Chwee Daniel
Associate Professor
Natural Sciences and Science Education
National Institute of Education
Nanyang Technological University

Preface

This curriculum writing project is a bold move, but one that is aligned to broader local and international science educators' efforts to forge stronger connections between science and learners' everyday life. In this book, we take the view that scientific practices are cultural practices and hence, scientific knowledge is cultural knowledge. As science educators, we hope that science teaching may be culturally responsive to students' personal experiences and interests.

The contributors of this book are all Singapore science teachers who share a common belief that we can make science more relevant to our students' lives by drawing upon elements from our unique local culture (e.g., food, clothes, sports) to pique students' interests and connect them to science concepts. We do this by *consciously* foregrounding the cultural aspect of the topic before introducing the science concepts.

We hope that this book will serve as a good resource for science teachers in Singapore and possibly other countries who may share some aspects of our culture.

Coeditors

Dr. TEO Tang Wee
Assistant Professor
National Institute of Education

Mr. KHOH Rong Lun
Chemistry Teacher
Temasek Junior College

Authors' Biographies

BOO Michelle is teaching Biology at CHIJ St Nicholas Girls' School (Secondary). Her keen interest in Biology began when she was first introduced to the subject at secondary three. She continued to study Biology in Junior College and eventually went on to pursue a major in Life Sciences (with specialisation in Biomedical Science) and a minor in Forensic Science at the National University of Singapore. She also spent one semester at the Université Pierre et Marie Curie in Paris and took Biology courses including Paleontology which was taught in French. Now a Biology teacher, Michelle is continually exploring activities to share her love of the subject with her students. She often include snippets of recent discoveries in science in her lessons as she believes that all learners of science should keep themselves abreast of the latest developments. In her free time, Michelle likes to play the piano or strum on her ukulele.
Email: michelle_boo_wee_siew@moe.edu.sg

CHEW Shuhui Eunice is a Mathematics and Chemistry teacher at Chung Cheng High School (Main). She graduated with a Bachelor's degree with honours in Applied Mathematics from the National University of Singapore. During her undergraduate studies, she went on a student exchange at the University of California, Santa Barbara and the University of California, Los Angeles. She graduated from the Postgraduate Diploma in Education programme at the National Institute of Education with distinction.
Email: chew_shuhui_eunice@moe.edu.sg

CHO Wen Jing majored in Chemistry at the University of Warwick in the United Kingdom, where she obtained her Bachelor of Science (Honours). Since her first teaching stint, her goal has been to try and make science interesting and relevant for students such that they would be able to see and appreciate the beauty in the world around them, as well as grow in their zeal for learning that is not just limited to science. It is her hope that they will ultimately become passionate individuals with a purpose in life. She hopes that the book will help teachers and students alike to discover and see things in ways they might not have seen before. Her other interests include sports, music, and photography. She is currently teaching at Nanyang Girls' High School.
Email: cho_wen_jing@moe.edu.sg

GAN Ghim Kui graduated from the National University of Singapore with Bachelor of Science (Honours) and majored in Chemistry. Teaching is his lifelong ambition as it is his love to see the students grow to become strong individuals with character who can contribute to the society and also, to the world. He is very intrigued by the aspect of critical and creative thinking in education and strives to design his lessons creatively for his students. His adage: "creative and critical thinking will be imperative for the nation to breach to the next level of maturity for the economy and society" is the driving force in his teaching career. Therefore, he hopes that his contribution to this book will be helpful for teachers to create a fun teaching and learning environment around the world. His other interests include philosophy, poetry, and sports. He is currently teaching at St. Andrew's Secondary School in Singapore.
Email: gan_ghim_kui@moe.edu.sg

KHOH Rong Lun is a Junior College Chemistry teacher. He graduated from National University of Singapore (NUS) with Bachelor of Science (Honours) in 2009, and was conferred Master of Science from NUS and Distinction for his Postgraduate Diploma in Education from National Institute of Education in 2012. Rong Lun enjoys being engaged in both science and education research work and is always excited to share the research skills he acquired with his students. He believes that science education is never complete without practical work and that educators should make use of inquiry-based experiments to stoke students' interests in science. Other than work, Rong Lun finds time to enjoy performing arts and movies, and to also volunteer at organisations like National Youth Council where he gets additional opportunities to work with more youths.

Email: khoh_rong_lun@moe.edu.sg

KOH Bing Qin is a Chemistry teacher at Pasir Ris Secondary School. He was actively involved in research work during his undergraduate days and is keen to share his passion in experimental work with his students. He believes that science should be taught through experiential learning and inculcating an inquiry mind should start at a young age.

Email: koh_bing_qin@moe.edu.sg

LIM Jia Ying Jessica is currently teaching at Woodlands Secondary School. She graduated from the National University of Singapore with a Bachelor of Science (Honours), majoring in Chemistry. She is passionate about bringing life to science both in and outside the classroom. She believes that by helping her students to embrace and understand Science, her students are able to open new doors to their future due to the strong emphasis on scientific knowledge in the world today. As a student, she has actively participated in many school and national Science fair activities that have led her to place paramount importance to thinking and learning outside the box. She hopes that this book is able to convey her passion for learning and teaching Science creatively. Her other interests include martial arts, sports, and reading.

Email: jessica_lim_jia_yin@moe.edu.sg

LIM Shan Yan is a Science teacher at Changkat Changi Secondary School. She earned a Bachelor of Science in Life Sciences (Honours) from the National University of Singapore. She also spent a semester at McGill University where she took interesting modules such as *Our Evolving Universe* and *Neural Basis of Behaviour*. She enjoys teaching Science at her school. She is thankful for the ample opportunities provided by her school to support her quest for nurturing budding scientists

or even simply getting her students to think about Science in ways they have never thought before.
Email: lim_shan_yan@moe.edu.sg

LIN Jiansheng is a beginning teacher at Nanyang Junior College and he teaches Chemistry. He graduated from the National University of Singapore with a Bachelor of Science (Honours). He enjoys jogging and swimming to unwind after a long day of school. During his leisure time, he also likes to try out new food where he encountered all kinds of flavours and was intrigued by the use of spices in Indian cuisine.
Email: lin_jiansheng@moe.edu.sg

LOW Wei Chuan Matthias graduated from Nanyang Technological University with a Bachelor Degree (Honours) in Chemistry and Biological Chemistry. He is a Chemistry teacher in a local secondary school. He is passionate about making Science come alive for students. He believes that teaching is about knowing the needs of students and crafting lessons to meet their learning needs. Reflecting upon his own teaching experiences, he incorporates localised cultural elements that resonate with students' experiences to help them connect with scientific theory and heighten their interest in Science.
Email: low_wei_chuan@moe.edu.sg

NG Shi Han earned a Bachelor of Science (Chemistry) degree with Honours from National University of Singapore. He recently graduated from the National Institute of Education with a Postgraduate Diploma in Education. Shi Han is currently teaching Chemistry and Mathematics at Assumption English School. He enjoys the daily challenges of educating and working with youths and believes in the cultivation of good values. He is the co-director of Save That Pen, a non-profit environmental organisation, which aims to reuse, reduce, and recycle pens by collection, sorting, refilling, and giving out to various beneficiaries. Through this organisation, he is able to educate students, as well as the public about environmental sustainability. Save That Pen (www.savethatpen.com), which envisions zero waste, has been engaging youths actively through their quarterly sorting sessions in University Town of National University

of Singapore to become green ambassadors, who will spread the green message to their friends and families. Shi Han believes in the use of real life examples to teach Science concepts and values, as it creates an impact and allow students to learn better.
Email: ng_shi_han@moe.edu.sg

SRINIVASAN Shyam received his Bachelor's and Master's degrees from United Kingdom and United States respectively. He is a passionate and committed teacher who is currently teaching Mathematics and Physics at Marsiling Secondary School. Apart from his focus on pedagogy as an effective platform in helping students in their learning, Shyam has his students' interest at heart and is equally committed to their character development. Besides Shyam's passion in nurturing his students, he also professes his deep interest in reading, computer programming, sports, and most importantly, chilling out with his friends.
Email: srinivasan_shyam@moe.edu.sg

TAN Yong Leng Kelvin taught Chemistry at a Junior College on weekdays and explored the reactivity of transition metal clusters at the National University of Singapore on weekends and school vacations. He continued to pursue his passion in research by taking a sabbatical year from teaching to delve into the realm of bioorganometallic chemistry at the École Nationale Supérieure de Chimie de Paris in France. More recently, he was a visiting scientist at the University of Neuchatel in Switzerland. Kelvin is currently a School Scientist at his alma mater, Hwa Chong Institution, where he collaborates with fellow science educators to mentor students in research projects, as well as work with external partners to organise Science-related activities.
Email: tanyl@hci.edu.sg

TANG Chi Sin is a committed Condensed Matter Physicist and is teaching Physics at NUS High School for Mathematics and Science for years 5 and 6, which he sees as an opportunity to induct impressionable young minds into the wonders of his discipline. When not running scattering experiments, crunching data or entertaining his students with his unique brand of humour, Chi Sin can be found swimming or doing long-distance running.
Email: nhstcs@nus.edu.sg

TEO Tang Wee is an Assistant Professor at the National Institute of Education, Nanyang Technological University. She pursued her doctorate and graduated from the University of Illinois, Urbana-Champaign in 2011. She was a former Chemistry teacher at Meridian Junior College and the National University of Singapore High School of Mathematics and Science. Currently, she teaches the Postgraduate Diploma in Education (Secondary and Junior College), Bachelor of Science (Education), and higher degree programmes at NIE. Her research focuses on cultural and sociological studies in Science education. She believes that making Science relevant to students' everyday lives requires the valuing of culture and foregrounding it in Science teaching.

Email: tangwee.teo@nie.edu.sg

Introduction

Teo Tang Wee and Khoh Rong Lun

Purpose of This Book

This book provides a platform for Singapore Science educators to engage in the curriculum design, innovation, and writing of culturally relevant Science activities for sharing with other Science educators locally and globally. There are two main goals of this book. First, it addresses the absence of a science resource book that values the cultural elements embedded within the highly cosmopolitan Singapore society. Through consciously integrating cultural elements into the design of science activities that Singapore teachers can use in their classrooms, the editors and authors of this book hope to invoke students' interest to learn Science through making connections to real life experiences that surround them in their everyday lives and matter to them. Second, it offers an educative opportunity for teachers to cooperate and collaborate professionally and intellectually with one another as they become a producer of knowledge in the making of a culturally relevant science curriculum. More importantly, it offers an avenue for Science teachers to have a voice and stake in the curriculum making of our local Science curriculum.

This book is an inaugural and purposeful effort to introduce culturally relevant science into the Singapore Science curriculum. Hopefully, it may be of interest to regional and international educators interested to enact their own personalised culturally relevant Science. Additionally, this book offers a lens into the cultures of Singapore and how her everyday functions, operations, and living are infused with Science. We encourage Science teachers in Singapore and elsewhere to adopt and adapt the ideas in this book to cater to their situated needs in their formal and informal Science teaching.

Contributors of This Book

This book encapsulates the efforts of science educators, current school science teachers and a science education researcher, who share a common belief that science should be taught in culturally relevant ways to enhance the relevancy, responsiveness, and currency of science teaching and learning. These teachers have various academic qualifications in science and teaching experiences. They belong to various age, gender, racial, and ethnic groups. The diversity of life experiences, cultural and scientific knowledge, and skills they bring to the writing of the book chapters add colours to science teaching and learning. At the same time, students simultaneously construct new knowledge and understanding about Singapore's history, culture, and society as they learn science. Included in this book are the authors' biography and reflections in writing the book chapter. These capture the diversity of experiences, qualities, interests, personal beliefs, and inspirations they bring to the writing of the activity. The book chapters thus, reflect the richness of teacher embodiments that can possibly be harnessed to advance the quality of teacher education, science teaching, and learning. We hope that following this book, more teachers would come forth to write and enact culturally relevant science, as we believe that this is a step forward to achieving a scientifically literate science citizenry.

Culturally Relevant Science

To teach science in culturally relevant ways means to integrate the cultural knowledge and practices of a local community with the Western scientific knowledge systems and practices to enhance the relevancy and applicability of science in our everyday lives. This entails the recognition of the diverse knowledge, perspective, beliefs, and practices of individuals and communities and infusing these with the canons of scientific knowledge so that a situated constructed knowledge of science is synergistically produced.

The science curricula in many countries are generally Eurocentric in nature and disengaged from the local elements of a community, society, or country. To date, there is no existing book in the Southeast Asian market and beyond which explicitly incorporates the cultural elements of this society. Most science resource books are also written for the general Western classrooms and the authors have limited understanding of the Asian contexts and characteristics. The imbuement of cultures into Science is hence, severely lacking and it is no wonder that students will feel that science is detached from their everyday lives and lose interests in science learning over time.

This book offers a culturally enriched science resource for teachers to enact culturally relevant science teaching. Our purposeful design of culturally relevant science activities acknowledges and validates what learners already know and build upon that knowledge toward a more sophisticated understanding of the local and Western perspective of science. The learners' in-school and out-of-school experiences could thus be bridged (Howard, 2003). The activities will also pique students' interests to conduct scientific inquiry, gain insights into other cultures, and foster collaboration and mutual understanding among learners from diverse cultural backgrounds. Aligned with this goal to promote teacher agency, the book chapters will mostly be written by teachers (rather than 'for' teachers) who are current practitioners to showcase their curriculum writing capabilities. In the final section of the book we will capture teacher reflections on how their experience with curriculum writing of culturally relevant science has reshaped their understanding of science teaching and learning. As such, each chapter is embedded with the funds of knowledge that each author brings to the construction of the chapter. In the enactment of the activities, teachers and students bring their funds of knowledge to the enactment of each activity hence, value-adding to their teaching and learning of science.

Funds of Knowledge

Science educators have argued that school science, when disconnected from students' home and community life may invoke feelings about science as impractical, alien, and contradictory to their beliefs and everyday practices (Bouillion and Gomez, 2001; Brickhouse, 1994). As such, students become less interested in science. Several researchers (Atwater, 1996; Hammond, 2001) have argued for science to be multicultural by grounding science instruction in students' cultural knowledge and experience. Premised on the belief that people are competent and develop knowledge derived from life experiences, Gonzales and Moll (2002) argue that the funds of knowledge students bring into the classroom to bridge their personal and school/classroom experiences must be valued as an epistemological construct of the classroom discourse. These funds of knowledge are rooted in practice and embody the culturally constructed knowledge, action, and disposition (habitus) (Basu and Barton, 2007). When students feel that they are able to connect and utilise their funds of knowledge in engaging ways, they will feel empowered to learn and participate, rather than resist, in investigative science learning (Bouillion and Gomez, 2001).

Significance of This Book

There are two main unique and significant features about this book.

First, this book is intended to empower science teachers to carry out science instruction in culturally relevant ways. In particular, the science activities are intentionally written in a non-prescriptive manner so that teachers are encouraged to explore, improvise, and adapt the activities according to their situated needs and contexts. We hope that other teachers will be inspired to think about and write other culturally relevant science activities.

Second, the ideas in the book will remain current for a long time. "Cultures", according to Clifford Geertz, a renowned anthropologist, is "a system of inherited conceptions expressed in symbolic forms by means of which men communicate, perpetuate, and develop their knowledge about and attitudes toward life" (1973, p. 89). This implies that learning science through a cultural lens will generate long lasting knowledge and understanding of science. Furthermore, the nature of the book content seeks to kindle awareness for the beauty of science as a form of common language in bridging cross-cultural ties, whereby the different culturally influenced activities, be it Eurocentric or Asian-centric, could be explained via the same basic principles of science.

In addition, the science activities will use common food consumed and materials used in Singapore so that learners could make connections to science. The real-life contexts of the problems would lend authenticity (Rahm *et al.*, 2003) to the tasks and hence enable students' learning could be more meaningful. The knowledge and skills acquired will thus, not expire. We would like to encourage teachers to engage in culturally-relevant science instruction as this, we believe, is a step toward achieving a scientifically literate citizenry in the long run.

In sum, our long term and continued goal is to accumulate more culturally relevant science activities over time. We hope that this book will inspire more teachers to write and enact culturally relevant science in their classrooms and beyond.

References

Atwater, M. (1996). Social Constructivism: Infusion into the Multicultural Science Education Research Agenda. *Journal of Research in Science Teaching,* 33, 821–837.

Basu, S. J., Barton, A. C. (2007). Developing a Sustained Interest in Science Among Urban Minority Youth. *Journal of Research in Science Teaching,* 44, 466–489.

Bouillion, L. M., Gomez, L. M. (2001). Connecting School and Community with Science Learning: Real World Problems and School-Community Partnerships as Contextual Scaffolds. *Journal of Research in Science Teaching,* 38, 878–889.

Brickhouse, N.-W. (1994). Bringing in the Outsiders: Reshaping the Sciences of the Future. *Journal of Curriculum Studies,* 26, 401–406.

Geertz, C. (1973). *The Interpretation of Cultures: Selected Essays.* New York: Basic.

Gonzales, N., Moll, L. C. (2001). *Cruzando el Puente*: Building Bridges to Funds of Knowledge. *Educational Policy,* 16, 623–641.

Hammond, L. (2001). Notes from California: An Anthropological Approach to Urban Science Education for Language Minority Families. *Journal of Research in Science Teaching,* 38, 983–999.

Howard, T. C. (2003). Culturally Relevant Pedagogy: Ingredients for Critical Teacher Reflection. *Theory Into Practice*, 42, 195–202.

Rahm, J., Miller, H. C., Hartlet, L., Moore, J. C. (2003). The Value of an Emergent Notion of Authenticity: Examples from Two Student/Teacher–Scientist Partnership Programs. *Journal of Research in Science Teaching*, 40(8), 737–756.

Chapter 1

Red Dragon Fruit: Using Red Pigment Extracts from Pitayas as Natural Indicators

Teo Tang Wee

Interesting Facts

1. Dragon fruit is native to Mexico, Central America, and South America.
2. This fruit is now commonly cultivated in East Asia, and Southeast Asia including Malaysia, Vietnam, Taiwan, the Philippines, and Thailand.
3. The plant can now be found in Japan, Hawaii, Israel, northern Australia, and southern China.

The Fruit of Dragons?

Pitaya is the fruit that grows on a red skinned climbing cactus known as *hylocereus* (Figure 1). This plant produces large white night blooming flowers known as "moonflower" or "Queen of the Night" and has a light melon taste. Pitaya is also commonly known as 龍珠果 (dragon pearl fruit) or 火龍果 (fire dragon fruit) in Chinese, *thanh long* in Vietnamese, *buah naga* in Malay, and *kaeo mangkon* in Thai. It has a beautiful bright red, leathery, and slightly leafy skin. The flesh of the fruit is translucent with tiny black seeds.

Figure 1.   Dragon fruits sold at a fruit stall.

An excerpt from an online forum

I am very worried after seeing red urine and bowel movements this morning. It happened after I ate two pieces of red flesh dragon fruit last night. Should I go see the doctor? Does anyone have experience with this? Please share.

Relevant to Our Everyday Lives

There are generally three kinds of pitayas — red skin with white flesh, red skin with red flesh (Figure 2), and yellow skin with white flesh. All three types, in particular, the red skin dragon fruits are imported into Singapore and can be found in the local wet markets, supermarkets, and fruit stalls. It has gained increasing popularity as it is known to contain antioxidants. As the quality of education improves and people become more scientifically literate and health conscious, the choice of food beneficial for health becomes more popularly consumed.

Figure 2.   The bright red colours of red dragon fruit flesh.

Figure 3.   Hylocereus undatus.

Most fruits in Singapore are not grown locally but imported from other countries (Figure 3 shows a dragon fruit plant). Established in April 1, 2000, the Agri-Food and Veterinary Authority of Singapore (AVA) is the national authority on food safety for both primary and processed food. Among its many responsibilities, it overlooks the export and import of meat, meat products, fruits, and vegetables. AVA conducts routine laboratory testing of food samples, for example, to check for excessive pesticide residues. A fine and suspension will be imposed on the importer if the consignment fails to meet AVA's regulated guidelines.

The red colouring in red beetroot is a commercial source of betacyanins — a type of water-soluble pigment. However, red beetroot contains geosmin and pyrazines that are responsible for the unpleasant peatiness of this crop and the high concentration of nitrates which can form carcinogenic nitrosamines (Phebe *et al.*, 2009). As such, the red pigment in red dragon fruit may possibly be extracted and used as red colouring instead as it does not give the same negative sensorial feel.

Can the Pigment in Red Dragon Fruit Flesh be used as a Natural Indicator?

Chemical composition

The colour of dragon fruit red flesh is attributed to betacyanins, which is a class of water-soluble pigments (Wybraniec *et al.*, 2007). Betacyanins are also found in red beetroots. The red pigments in red cabbages, often used to illustrate natural indicators in school-based experiments, are however, anthocyanins.

Betacyanins and anthocyanins have different chemical structures although both are water-soluble. With this knowledge, how would you address the anonymous writer's concerns (on page 9)?

Guiding question

Red cabbage pigments can be extracted as natural indicators to test for acidity and alkalinity differences. Can the pigment in dragon fruit red flesh serve the same function?

Extraction procedures

Peel a red flesh dragon fruit and cut up the flesh into small chunks. Put the cut up fruit into a beaker and add enough water to cover the flesh (Figure 4). Stir the mixture to extract as much red pigment as possible. When the solution is sufficiently red in colour, filter the colouring into a conical flask using a filter funnel and filter paper (Figure 5).

Figure 4.   Cut up dragon fruit flesh mixed in water.

Figure 5.   Filtering the mixture to extract the red pigment.

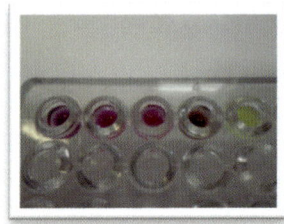

Figure 6. Colours of the red dragon fruit extract in pH 1, 2, 7, 10, and 13 (from left to right).

Testing pH changes

The extract was added dropwise to solutions with various pH (1, 2, 7, 10, 13) and obvious colour change was observed at pH 10 (Figure 6). This shows that the red pigment in red flesh dragon fruit can be used as a natural pH indicator at high pH. What can you deduce about the pH of urine and bowel movements of the anonymous writer?

Student Inquiry

Suggested questions for student research

•   What can be some possible commercial uses of the extracted red pigments from dragon fruit?

- How can the red pigment be extracted and packaged for commercial sale?
- What are some other fruits or vegetables that allow for colour pigments to be extracted and used as natural food indicators?
- What are the advantages and disadvantages of using natural indicators as compared to synthetic indicators?
- How are betacyanins and anthocyanins similar and different e.g., in terms of chemical structures?
- Why is the red colour pigment in red dragon fruit water-soluble?
- What happens to the chemical structure when it changes colour under different pH conditions?
- How would the titration results using the red pigment from dragon fruit flesh differ from synthetic pH indicators such as phenolphthalein and methyl orange?

Author's note

- *This activity is considered low-cost as two dragon fruit would yield sufficient red colouring extract for a class of 30 students. As students learn science, they will also learn not to waste food and resources. Cultivating good habits as such is also part and parcel of science teaching and learning.*

References

Phebe, D., Chew, M. K., Suraini, A. A., Lai, O. M., Janna, O. A. (2009). Red-Fleshed Pitaya (*Hylocereus polyrhizus*) Fruit Colour and Betacynanin Content Depend on Maturity. *International Food Research Journal*, 16, 233–242.

Wybraniec, S., Nowak-Wydra, B., Mitka, K., Kowalski, P., Mizrahi, Y. (2007). Minor Betalains in Fruits of *Hylocereus* Species. *Phytochemistry*, 68, 251–259.

Chapter 2

Delectable Blue!

Koh Bing Qin

Nyonya Kuehs (bite-size Peranakan confectioneries) are very popular for tea and breakfast in Singapore and some neighbouring countries. Apart from their fragrance and great taste, their appealing colour is another reason for their popularity. It often leaves us to wonder: What contributes to these attractive colourings? My grandparents often said that the colourings are healthy as they are extracted from natural sources. While this may be true in the past, it may not be the case in today's food industry due to the high production costs and time constraint in making these food items for sale. One may be curious to find out if the food dyes used in these confectioneries are from natural sources, man-made or both.

Let us investigate the sources of blue colouring in one of our all-time favourite *Pulut Inti*, which has traditionally been dyed with butterfly pea (or *Clitoria ternatea*) flower, also known as *Bunga Telang*. This flower is common in tropical regions including Southeast Asia (Suebkhampet and Sotthibandhu, 2012). In addition, we will also attempt to extract the blue colouring from butterfly pea flowers and test it as a pH indicator. Through these activities, one can learn about natural food dyes — which have less health side effects than synthetic food dyes — in local traditional food, and learn that it is not true that all blue food colourants are "unnatural".

Part I — Identifying the Source of Colourant

The objective of the science activity

The first part of the chapter aims at illustrating simple chromatography techniques used to separate colourings in food.

Description of the science activity

This activity involves the extraction of the blue colourant from *pulut inti* and comparing it to the extracted colourant from butterfly pea flowers using thin layer chromatography. The materials and instruments required for this activity include:

1. 0.5–1 *pulut inti*
2. Dried butterfly pea flower
3. Approximately 80% ethanol in water
4. Beakers and conical flasks
5. Hot plate heater
6. Retort stand with clamp
7. Silica coated chromatography sheet

The colourant from dried butterfly pea flowers can be easily extracted by soaking the flowers in water for at least an hour (Figure 1). About seven petals and a teaspoon of water (5–6 cm^3) are sufficient to ensure a good concentration. Pound the petals gently with a stirring rod to extract more colourant.

Figure 1.   Extraction of blue colourant from butterfly pea flowers using water.

Figure 2.   Rinsing *pulut inti* with ethanol (left). The rinsed *pulut inti* in a dry conical flask (right).

The blue colourant in *pulut inti* can be extracted using ethanol. Place the *pulut inti* (without the coconut toppings) in a beaker. Rinse the *pulut inti* with some ethanol a few times to remove as much oil and coconut milk as possible (Figure 2). The amount of ethanol should be enough to cover the *pulut inti* in the beaker. While rinsing, attempt to break the glutinous rice into smaller pieces using a stirring rod. The rice should turn harder after rinsing. Drain the ethanol and transfer the glutinous rice to a dry conical flask.

B.Q. Koh

Add ethanol to the conical flask with the rinsed glutinous rice. The volume of ethanol should be just enough to submerge the glutinous rice and an excessive amount of solvent should be avoided. Place the conical flask in a boiling bath to evaporate most of the ethanol (Figure 3). Stir the content in the conical flask occasionally.

When the amount of ethanol in the conical flask is reduced to about half its original amount transfer the remaining ethanol into a smaller beaker or conical flask using filtration. The colour of the solution should be pale blue at this stage. Evaporate the remaining ethanol to almost complete dryness (Figure 4).

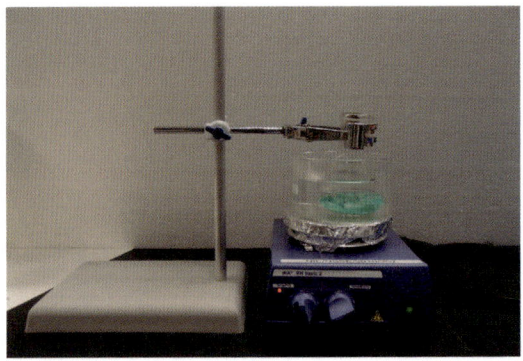

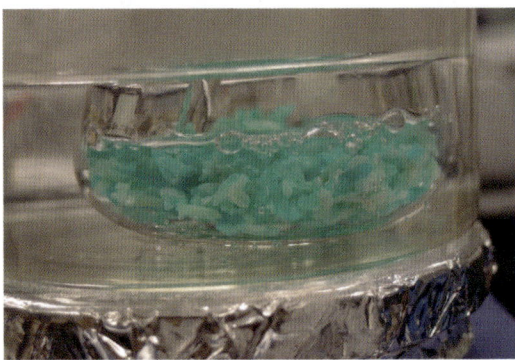

Figure 3.   Experimental set-up to extract the blue colourant (top). A close-up view of the glutinous rice (submerged in ethanol) being heated.

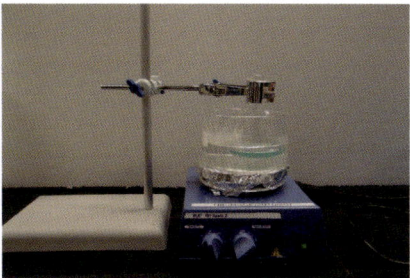 

Figure 4.   Concentrating the extract (left). The concentrated blue colourant (right).

Figure 5.   Comparison of *pulut inti* extract (**P**), butterfly pea extract (**F**) and a co-spot (**C**) consisting of **P** and **F**.

With both the extract from the butterfly pea flower and *pulut inti*, students can now find out the type of colourants in *pulut inti*. This can be done using thin layer chromatography separation techniques. A silica coated chromatography sheet and a solvent system of 50% ethanol in water can be used. Figure 5 shows the chromatogram using this solvent system. The results show that the colourant from the tested *pulut inti* differed from the blue colourant in butterfly pea flower.

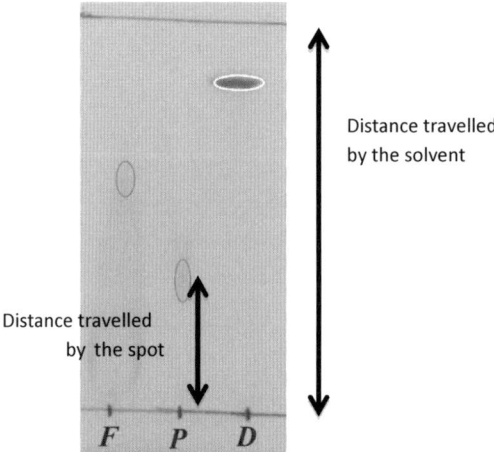

Distance travelled
by the solvent

Distance travelled
by the spot

F   P   D

Figure 6.   Comparison of *pulut inti* extract (**P**), butterfly pea extract (**F**) and Brilliant Blue
E133 food dye (**D**).

Apart from dried butterfly pea flowers, students may compare the
pulut inti colour extract with the colourants found in other kinds of
food such as blueberry jam, blackcurrant syrup, and food dyes.
Figure 6 shows a thin layer chromatogram that compared *pulut inti*
extract, butterfly pea flower extract and diluted Brilliant Blue E133
food dye.

Other than making qualitative observations, one can calculate the
retention factor (R_f) to determine the identity of the colourants found in
pulut inti.

Retention factor (R_f) = distance travelled by the spot ÷ distance travelled
by the solvent.

Inquiry learning for students

1. Simple paper chromatography is only one type of chromatography
    technique. What other methods are there to determine the identity of
    the food colourants?

Quick facts 1

While blue is the favourite colour of many people, it is widely believed that the colour has one of the greatest ability to suppress appetite. The aversion to this popular colour on the plate is plausibly because it appears less often as natural food colour (Morton, 1995). Apart from blueberries and butterfly pea flower, examples of edible blue products seem rare. Furthermore, as the colour is often associated with toxic materials and food spoilage (Magoulas, 2009), we tend to consciously or subconsciously reject the colour in our diet to avoid being poisoned.

Inquiry learning for students

1. Do you agree with the above claim?
2. If such a consumer preference is true, then what is the reason for confectioneries and factories to continue producing blue food products such as blue coated chocolates, sweets, and drinks?

Quick facts 2

Not all natural blue colourants are lethal. In fact, some blue plant extracts have been known for their medicinal and ornamental values. For instance, butterfly pea flowers have been known to possess antioxidant properties (Umesha *et al.*, 2012). Their roots are also used by traditional healers in Tamil Nadu to treat indigestion, headache, and eye diseases (Muthu *et al.*, 2006). More recently, butterfly pea flower extract is used as a dye on animal blood smear staining (Suebkhampet and Sotthibandhu, 2012) and was noted for its potential, in combination with other plants, for the treatment of diabetes mellitus (Adisakwattana *et al.*, 2012).

Inquiry learning for students

1. What are some uses and benefits of natural colourants?
2. What are the chemical substances responsible for the positive health attributes described above?
3. What are the possible drawbacks of using natural colourants from plant sources?

Part II — Acids, Alkalis & pH Indicators

Quick facts 3

Students may be curious to find out what constitutes the blue colourants in butterfly pea flowers. The truth lies in the blue petal lines which consist of anthocyanin with glucosylation at the 3'- and 5'- positions (Kazuma *et al.*, 2003). Anthocyanin belongs to a parent class of compound known as flavonoids which are widely found in plants. Anthocyanins are water-soluble pigments which appear as different colours at different pH (Wrolstad, 1993). This fact allows us to make use of the extract from butterfly pea flowers as a pH indicator. Apart from butterfly pea flower, other plant products rich in anthocyanins include red cabbage, red dragon fruit, blueberry, cranberry, raspberry, and blackcurrant.

Suggested science activity

I. Illustrating the pH indicator property of butterfly pea flower extracts.

    Different solutions of pH 1 to pH 13 can be prepared using aqueous hydrochloric acid and sodium hydroxide solutions. The butterfly pea flower extract can be obtained using the same method described earlier. A simple filtration can be performed to remove the flower petals from the aqueous extract.

    Add about five drops of the flower extract to test-tubes containing approximately 3 mL of aqueous hydrochloric acid or sodium hydroxide solution. Different colours may be observed at different pH (see Figure 7).

II. With the colours of the flower extract at different pH identified, the mixture can now be used to predict and determine the pH of different household products such as vinegar, fruit juices, soft drinks, and soap.

Quick facts 4

Aside from wild *Clitoria ternatea*, various colour mutants of the flower are known including the white and mauve flower mutants (Kazuma *et al.*, 2003).

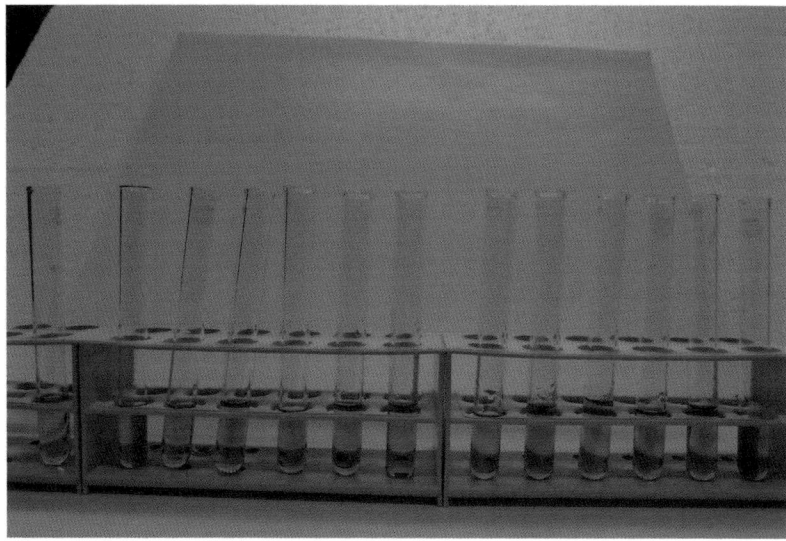

Figure 7.   The colour of butterfly pea flower extract in solutions with pH ranging from 1 (red on the left) to 13 (green on the right).

Inquiry learning for students

1. What are the differences in the chemical content of the petal lines of different coloured flowers?
2. Do you expect the other flower petals to contain colourants that can act as pH indicators? How can you test your idea?

Notes for users

Several solvents commonly available in school science laboratories have been used to extract the colourant from *pulut inti*. It has been found that ethanol is the most effective solvent. However, ethanol is flammable and should not come into direct contact with a naked flame. Where possible, the evaporation of the solvent should be done in the fume cupboard using a hot plate.

References

Adisakwattana, S., Ruengsamran, T., Kampa, P., Sompong W. (2012). *In Vitro* Inhibitory Effects of Plant-Based Foods and Their Combinations on Intestinal α-Glucosidase and Pancreatic α-Amylase. *BMC Complementary and Alternative Medicine*, 12, 110.

Kazuma, K., Noda, N., Suzuki, M. (2003). Flavonoid Composition Related to Petal Color in Different Lines of *Clitoria Ternatea*. *Phytochemistry*, 64, 1133–1139.

Magoulas, C. (2009). How Color Affects Food Choices. *UNLV Theses/ Dissertations/Professional Papers/Capstones, Paper 552*, University of Nevada, Las Vegas, NV.

Morton, J. L. (1995). Color and Appetite Matters. *Color Matters*. Available at http://www.colormatters.com/color-and-the-body/color-and-appetite-matters [accessed on December 11, 2012].

Muthu, C., Ayyanar, M., Raja, N., Ignacimuthu, S. (2006). Medicinal Plants Used by Traditional Healers in Kancheepuram District of Tamil Nadu, India. *Journal of Ethnobiology and Ethnomedicine*, 2, 43.

Suebkhampet, A., Sotthibandhu, P. (2012). Effect of Using Aqueous Crude Extract from Butterfly Pea Flowers (*Clitoria Ternatea L.*) as a Dye on Animal Blood Smear Staining. *Suranaree Journal of Science and Technology*, 19, 15–19.

Umesha, S., Marahel, S., Aberomand, M. (2012). Antioxidant and Antidiabetic Activities of Medicinal Plants: A Short Review. *Asian Pacific Journal of Tropical Biomedicine*, 1–12.

Wrolstad, R. E. (1993). *Color and Pigment Analyses in Fruit Products*. Oregon: Agricultural Experiment Station, Oregon State University.

Chapter 3

Turmeric Spice as Natural Dye

Lin Jiansheng

Biriyani rice, masala curry, and tandoori chicken. These may be very familiar local Indian cuisine to Singaporeans and what do they have in common? Found inside all these dishes, lies a deep orange-yellow spice, turmeric which is used extensively in South Asian and Middle Eastern cuisine for its rich colour and fragrance (Prance and Nesbitt, 2005). Turmeric, also known as *Curcuma longa* has also been traditionally used for the treatment of inflammation, skin wounds, and removal of superfluous hair in Indian folk medicine.

Over the years, turmeric has been touted as "India's holy powder" with modern medical research highlighting its potential anticancer properties brought by the active ingredient, curcumin (Prasad and Aggarwal, 2011). Could it be a coincidence that the rate of cancer among Indians is lower than that of the Chinese and Malays in Singapore? (*Indian males: 138 per 100,000 population compared to Malay males 174 and Chinese males 328; Indian females: 164 per 100,000 population compared to Malay females 214 and Chinese females 330 in 2011*). While it is difficult to make comparisons in the case of Indians because of their small population and the relatively lower cancer incidence among Singapore Indians could be due to other factors such as more of them being vegetarians; researchers are also recognising the anticancer properties of turmeric given that the spice is an integral part of the Indians' diet (Health Promotion Board, 2013).

Turmeric is not used exclusively by the Indian community, it has also found its roots in China where it is widely used as a herbal medicine in traditional chinese medicine (TCM). Native to India, turmeric reached China by 700 AD during the Tang Dynasty when it was a time of great international trade. Known as *jiang huang* amongst the Chinese, it is used to promote the movement of "*qi*" (energy) for relief of epigastric and abdominal pain due to stagnant *qi*. Interestingly, turmeric also expels wind (an external cause of diseases brought by changing weather patterns) and promotes blood flow to treat arthritis, also known as wind-damp in TCM.

Like how the use of turmeric spread to other parts of the world for its medicinal uses, spices such as pepper, cloves, nutmeg, ginger, and cinnamon were also highly sought after (Czarra, 2009). Notably, Singapore's earlier history was centred on the trading of some of these spices. Between the late 13 and 14 centuries, Singapore then called as Temasek, functioned as a port-city where fleets of *prahu* (a type of sailing boat originating in Malaysia and Indonesia) carrying spices such as nutmeg and cloves from Moluccas gathered along with Chinese junks and Indian and Arab ships carrying their respective trade goods (Kwa *et al.*, 2009). Following which, the long-standing Anglo-Dutch competition for control of the lucrative spice trade led to the establishment of an East Indian Company station on Singapore by Sir Stamford Raffles on 1819, marking the beginning of modern Singapore history.

Objective of the Activity

Historically, turmeric has been used as a natural dye in the early 20th century to dye wool, silk, and cotton. When used with logwood, turmeric was used to dye ostrich feathers and skin rugs black. Due to its bright and clear yellow colour, turmeric is also used to colour food.

Turmeric is more than just a spice used for cooking and for its medicinal properties. Like many other plants, turmeric serves as the source of most natural dyes and different parts of these plants are used to produce dye colours (Dean, 2010). For example, fruits like blackberry, vegetables such as red cabbage, and tea leaves are extracted for their myriad of hues. While nature provides a readily available wealth of colours, the problem with natural dyes is that large quantities of natural materials had to be harvested to make small amounts of dye. Other concerns include the less uniformity and durability of natural dyes which imparts as compared to synthetic dyes, for instance, turmeric has a tendency to fade rapidly under sunlight and is sensitive to soap and other alkalis.

Given the needs of the textile industry to provide predictable and durable colours, characteristics that are important in the mass production of fabric and driven by economic factors, most modern commercial textile products are coloured with synthetic dyes which are chemically formulated in laboratories. Hence, the application of synthetic dyes grew rapidly and correspondingly the use of natural dyes began to decline. However, the rising environmental concerns globally have stimulated public interest in dyes that produce less contamination and manufacturers of textile products have started re-looking the use of natural dyes for dyeing.

This activity provides hands-on experience in dyeing and understanding the chemical interactions between the dye and the fabric.

Description of the Activity

The dyeing process involves several processes — preparing the material by cleaning, providing a mordant so that the dye will better adhere to the material, preparing a dye bath and immersing the material into the dye bath (Samanta and Konar, 2011).

Cleaning the material

Before dyeing, the textile fibres must be thoroughly cleaned and rinsed to remove grease or dirt particles; otherwise the dye colour may be applied unevenly. This treatment is sometimes referred to as "scouring". Cotton is cleaned by simmering. In the process, the various particles are released from the material into the water, causing a yellowish solution.

1. Weigh the material to be dyed and record the mass.
2. Place the material into a beaker and pour in enough water so that the material is well covered. If necessary, use a stirring rod to fully immerse the material.

Figure 1.   Heating of the cut up materials.

3. Heat the water to just below a simmer (i.e., boiling point of water) and keep the water at this temperature for approximately 30 minutes to an hour (see Figure 1).
4. Remove the material and rinse thoroughly in warm water until the water comes out clear.

Mordanting the material

Although many plant dyes may colour material when used without a mordant, the addition of mordant helps the dye to better adhere to the fibres, producing much stronger colours and increased light- and water-fastness. This mordanting process is often carried out before dyeing with the use of generally two types of mordants. Natural mordants include staghorn sumac and rhubarb leaves which are rich in tannin and oxalic acid respectively. A chemical mordant, known as alum is used here due to its versatility for different materials. Other alternatives include copper sulphate and ferrous sulphate.

1. Weigh out 20% alum or 4 teaspoons per 100 g of fibres.
2. Dissolve the alum in boiling water in a beaker with stirring to ensure well mixing.
3. Add the cleaned material into the beaker containing dissolved alum, making sure that it is fully submerged.
4. Allow it to sit in the mordanted water for 30 minutes to an hour, then remove the material and rinse with water.

Preparing a dye bath

This process involves the extraction of colour from plant materials. Most plant materials require simmering for some time in order to fully extract their colour. While some plant materials can be soaked in cold water, the use of hot water would help to speed up this process. After simmering for some time, the liquid is strained or filtered to remove undissolved powder or pieces of plant material in order to ensure even dyeing. As a general rule, use at least half and up to equal weights of dyestuff to fibre in order to produce reasonably strong colours. Taking note that the strength of the dye bath depends on the amount of dye colour present in relation to the amount of fibres added, adding more water would not dilute the dye colour.

1. Weigh out 50% turmeric powder or 10 teaspoons per 100 g of fibres.
2. In a beaker, mix the turmeric powder into a paste with warm water.
3. Add more water to the beaker with stirring to incorporate all the particles.
4. Bring the solution to a simmer and leave for about 30 minutes to an hour.
5. Strain or filter the liquid into another beaker (see Figure 2).

Dyeing the material

Many plant dyes can be applied without heat, however heating the dye bath will speed up the process for the colours to develop on the fibres. If the material is not still wet from being mordanted, it is necessary for the material to be dampened first. During simmering, move the material

Figure 2.   Filtration of the boiled mixture.

occasionally without agitating the bath excessively in the dye liquid to ensure even distribution of dye colour.

1. Place the mordanted material into the dye bath and ensure that the material is fully immersed by adding more water if necessary.
2. Heat the dye bath and simmer for 30 minutes to an hour.
3. Using a pair of tongs, remove the material from the dye bath.
4. Rinse the material in water until the water is clear.
5. Leave the material to dry.

Try tie-dyeing (see Figure 3) — a traditional technique which remains part of the rich culture of decorating fabric in Asia (Brown, 2001; Walter and Priestley, 2002). The basic method is to tie up the fabric with string, rubber bands, clips or even clothes pegs and the way you tie, fold or twist the fabric will determine the pattern obtained. The visual effect created is based on the fact that the dye is unable to thoroughly penetrate layers of tightly compressed fabric.

Notes for users

Many plants have poisonous parts and, similarly powdered dyestuffs, mordants can be harmful if ingested. Similarly, use separate equipments such as containers and strainers for mordanting and dyeing and do not use them for food preparation. Since there is heating involved, always be aware of hot surfaces and liquids that can scald or burn. Wear rubber gloves to

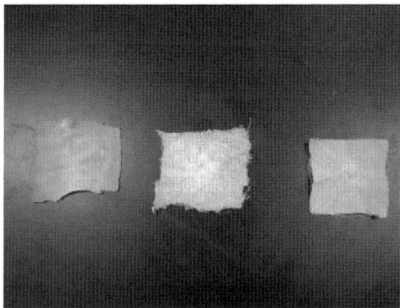

Figure 3.   Dyed cloth using tie-dyeing technique.

protect your skin and dust musk to avoid inhaling fine powders. Beware that the dyes can stain your clothes and always wipe down work surfaces well after dyeing. Lastly, dispose dyeing substances responsibly.

Inquiry Questions

1. How does turmeric powder impart its colour to the fabric? What is responsible for its orange-yellow colour?

   Answer:
   Colour is a natural phenomenon, the perception of which requires visible light — a full spectrum of colours comprising red, orange, yellow, green, blue, indigo, and violet in order of decreasing wavelengths (Brown, 2001; Walter and Priestley, 2002). When light strikes a surface, the surface absorbs some wavelengths and reflects others. During dyeing, colour-producing agents present in turmeric and other natural dyes, called chromophores are extracted and become attached to the textile fibres. These attached chromophores then determine which wavelengths will be absorbed and which other wavelengths that will be reflected, thus determining the perceived colour by the human eye.

Chemical structure of curcumin — chromophore in turmeric

These chromophores are able to impart colour due to the presence of conjugated systems (a structure with alternating double and single bonds) which exhibit resonance of electrons and impart stability to such compounds. Most dyes also contain groups known as auxochromes (colour helpers), examples include carboxylic acid, sulphonic acid, amino, and hydroxyl groups. Turmeric appears orange-yellow colour because it is absorbing red, green, blue, indigo, and violet wavelengths and reflecting orange-yellow to the human eye.

2. How does the addition of chemical mordant help the dye better adhere to the fibres?

Answer:
During dyeing, tiny molecules of dye matter stick to the fibres of the material and this sticking process is strengthened by mordants that increase the ability of the dye molecules to bond with the fibres (Vanlar, 2000). In the mordanting process, the fabric is treated with a metal salt solution (aluminium, chromium, copper, and iron salts). These metal salts anchoring to the fibres, attract the dye molecules to the fibres and creates the chemical bridge between the dye molecules and the fibres by forming co-ordinated complexes. The bond between the two involves the formation of covalent bonds or hydrogen bonds (can you identify them?) and other interactional forces in the structure below.

References and Additional Recommended Resources

Brown, P. (2001). *Decoration on Fabric*. Lewes: Guild of Master Craftsman.

Czarra, F. R. (2009). *Spices: A Global History*. London: Reaktion.

Dean, J. (2010). *Wild Colour: How to Grow, Prepare and Use Natural Plant Dyes*. London: Mitchell Beazley.

Health Promotion Board (2013). Singapore Cancer Registry Interim Annual Registry Report Trends in Cancer Incidence in Singapore 2007–2011. Singapore. Retrieved from http://hpb.gov.sg/HOPPortal/content/conn/HOPUCM/path/Contribution%20Folders/uploadedFiles/HPB_Online/Publications/.

Kassinger, R. (2003). *Dyes: From Sea Snails to Synthetics*. Brookfield: Twenty-First Century Books.

Kwa, C. G., Heng, D. T., Tan, T. Y., National Archives (Singapore) (2009). *Singapore, a 700-year History: From Early Emporium to World City*. Singapore: National Archives of Singapore.

Prance, G. T., Nesbitt, M. (2005). *The Cultural History of Plants*. New York: Routledge.

Prasad, S., Aggarwal, B. B. (2011). Turmeric, the Golden Spice in From Traditional Medicine to Modern Medicine. In *Herbal Medicine: Biomolecular and Clinical Aspects*, I. F. Benzie and S. Wachtel-Galor (eds.). Boca Raton: Taylor & Francis.

Samanta, A. K., Konar, A. (2011). Dyeing of Textiles with Natural Dyes, Natural Dyes. Retrieved from: http://www.intechopen.com/books/natural-dyes/dyeing-of-textiles-with-natural-dyes.

Vankar, P. S. (2000). Chemistry of Natural Dyes. *Resonance*.

Walter, C., Priestley, P. (2002). *The Basic Guide to Dyeing and Painting Fabric*. Iola, WI: Krause Publications.

Chapter 4

Bubble Tea Toppings

Chew Shuhui Eunice and Ng Shi Han

Bubble Tea

Culture in Singapore

Bubble tea is a popular beverage among Singaporeans and especially the youths. It is a common sight to see crowds of people queuing and waiting outside bubble tea shops. Bubble tea is commonly known as 泡泡茶 (pào pào chá); foam tea or 珍珠奶茶 (zhēn zhū nǎi chá); pearl milk tea. Over the years, there is an increasing variety of toppings that can be added to the bubble tea. This includes white pearl jelly, *Ai-yu* jelly, pudding jelly, *Azuki* beans, etc.

History

Bubble tea originated from Taiwan and was discovered in the early 1980s. There is no definitive history but it has been generally accepted that Ms. Lin Hsiu Hui invented bubble tea when she poured a typical Taiwanese dessert known as Fen Yuan, a sweetened tapioca pudding, into an ice tea drink.

What is bubble tea?

Bubble tea is a cold tea-base beverage usually infused with flavourings and/or milk. The mixture is shaken vigorously in a cocktail shaker to produce "bubbles", hence its name. An additional topping

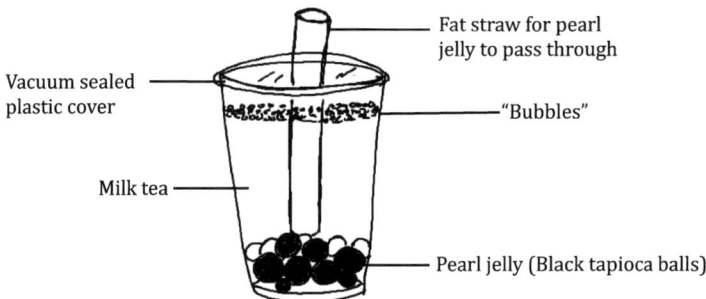

Vacuum sealed plastic cover — Fat straw for pearl jelly to pass through — "Bubbles" — Milk tea — Pearl jelly (Black tapioca balls)

known as pearl jelly (black tapioca balls) that looks like "bubbles" is subsequently added to it. Now, the term "bubble" refers not just to the "bubbles" formed on top of the drink but also the black tapioca balls at the bottom.

Class Activity

Some people are known to develop allergies toward starch. Can a person allergic to starch select any type of bubble tea toppings to be added to their bubble tea? What test can be conducted to verify if the bubble tea toppings have starch content?

Objective

This activity is to determine the presence of starch in five bubble tea toppings, namely pearl jelly, white pearl jelly, *Ai-yu* jelly, pudding jelly, and *Azuki* beans using a starch test.

Preliminary test for starch — the "chewing test"

Chew about five pearl jellies. How do the pearl jellies taste as you chew longer? Do you notice that it becomes sweeter over time? Why is that so?

Why does the pearl jelly taste sweeter as you chew it longer?

A: As you chew the pearl jellies longer, a secondary sugar known as maltose is produced, hence the sweeter taste. As hydrolysis of starch by salivary amylase enzyme produces maltose, this indicates the presence of starch in the pearl jellies.

Experiment

Apparatus and materials

1. Iodine solution of concentration 0.05 mol dm^{-3}
2. A bottle of distilled water

3. Small portions of bubble tea toppings:

    a. *Ai-yu* jelly
    b. Pudding jelly
    c. *Azuki* beans
        (or Japanese red bean)
    d. White pearl jelly
    e. Pearl jelly
    (Purchased from any
    bubble tea store.)
4. 5 Evaporating dish
5. 5 Plastic spatulas

Procedure

1. Wash five spatulas using distilled water.
2. Use a spatula to transfer small portions of bubble tea topping to an evaporating dish respectively as shown below:

Top: *Ai-yu jelly, white pearl jelly, pearl jelly* (*Left to Right*)

Bottom: *Azuki beans, pudding jelly* (*Left to Right*)

3. Wash the bubble tea toppings with distilled water.

> **Food for Thought:**
>
> Why do you need to wash the bubble tea toppings with distilled water?
>
> To remove any impurities that might affect the accuracy of the results that is based on the bubble tea toppings alone.

4. Add 3–5 drops of iodine solution to each of the bubble tea toppings. Observe any colour change. Record your observations for each bubble tea topping. The table below shows how you can write your observations.

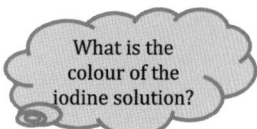

What is the colour of the iodine solution?

Presence of starch	Observations
Yes	Iodine solution formed a blue-black complex
No	No change; iodine solution remain brown

Teaching Tips: Below is the Predict–Explain–Observe–Explain (PEOE) approach which may be used with this activity. Below are some suggested questions to guide the PEOE process:

Predict: What would you expect to observe when the iodine solution is added to each bubble tea topping?

Explain: Why would you expect a similar or different colour change of iodine solution for the bubble tea toppings? What causes the iodine solution to change colour?

Observe: What is the colour change of the iodine solution when added to separate samples of the bubble tea topping?

Explain: What can you conclude from the observations?

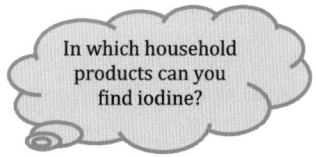

Note:

(a) The colour change of iodine with *Azuki* bean is not as apparent. Observe closely for any colour change.

(b) Consider meshing the bubble tea toppings for more accurate and clearer results.

Safety Precautions:

(a) Handle the iodine solution with care as it can stain the skin and clothing.

(b) Do not ingest the bubble tea toppings with iodine solution as iodine is poisonous.

What is starch made up of?

Q:

Starch is a carbohydrate that exists in two types of molecules.

1. Amylose (linear chain of glucose molecules):

2. Amylopectin (branched chain of glucose molecules):

The ratio of amylose and amylopectin varies according to the type of food.

Q: Iodine solution formed a blue-black complex in starch. What is the blue-black complex formed?

The iodine molecule slips inside the amylose coil forming a blue-black complex. Only starch containing amylose will give a darkblue complex as shown below.

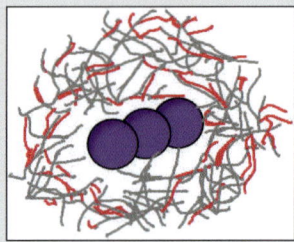

Observations and results

Presence of starch	Bubble tea toppings
	Pearl jelly and *Azuki* beans

Yes	
	Ai-yu jelly, white pearl jelly, and pudding jelly

No

Extension

In recent years, there were several reported cases of pearl jellies being recalled as they were suspected to contain traces of maleic acid that might cause kidney damage in the long term. Some bubble tea stores in Singapore had temporarily suspended the sale of bubble tea with pearl jelly toppings as a result.

Further Exploration — Case Study

Imagine that you have been tasked to investigate whether a packet of pearl jellies contains traces of maleic acid. Design an experiment to carry out this investigation.

The experimental design should include:

- Introduction including the description and uses of maleic acid, and the positive and negative health implications of maleic acid
- Objective of the experiment
- List of relevant apparatus and materials
- Detailed procedure
- Diagrams that show the experimental process and setup
- Table of recorded findings
- Conclusion
- Limitations of the experiment (e.g., possible sources of error).

Further Exploration — Geometric Isomerism

Maleic acid is a geometric isomer of fumaric acid. Compare the two molecular structures below.

Use balls and sticks to build a molecular model of the structures below. What do you think is "geometric isomerism"? What results in the formation of geometric isomers?

Maleic Acid   Fumaric Acid

Chapter 5

Milk Tea = Teh-C and Teh

Ng Shi Han and Chew Shuhui Eunice

Milk tea (see Figure 1) has always been a popular beverage among Singaporeans. It is a very common beverage found in "kopi tiams". A "kopi tiam" or coffee shop, a place where food and beverages are sold, is traditionally found in Southeast Asia. The word *kopi* is a Malay word for *coffee* (as borrowed and altered from the Portuguese) and *tiam* is the

Figure 1.   A picture of milk tea.

Hokkien dialect word for *shop* (店), hence the name *kopitiam*. The type of tea that is commonly sold in the kopitiams is black tea.

Tea plants are native to East and South Asia, and probably originated around the point of confluence of the lands of Northeast India, North Burma, and Southwest China. Although there are tales of tea's first use as a beverage, no one is sure of its exact origins. The first recorded drinking of tea is in China, with the earliest records of tea consumption dating back to the 10th century BC. It was already a common drink during the Qin Dynasty (3 BC) and became widely popular during the Tang Dynasty when it was spread to countries like Korea and Japan. The appreciation of tea has since then been spread to the Western nations and other parts of the world in the 19th century.

One needs to learn a diverse vocabulary before he/she can order a cup of tea in a coffee shop. "*Teh*" is the Hokkien dialect word for tea (茶). Sugar is usually added into the tea and the tea can be served with or without milk. "*Teh-O*" is tea without milk, while "*Teh*" is tea with condensed milk and "*Teh-C*" is tea with evaporated milk and sugar.

Table 1 shows the different local names of the tea beverage, with and without milk and sugar, sold in Singapore.

Table 1.   Different local names of tea beverages in Singapore.

Name	Description
Teh	tea with condensed milk
Teh-C	tea with evaporated milk and sugar
Teh-C-kosong	tea with milk and no sugar
Teh-O	tea with sugar only
Teh-O-kosong	tea without milk or sugar
Teh-peng	tea with ice, also known as *Teh*-ice
Teh-siu-dai	tea with milk and less sugar
Teh-gah-dai	tea with milk and more sugar
Teh tarik	the Malay tea described above
Teh halia	milk tea with ginger water

Objectives of the Activity

Have you wondered whether there are any science concepts related to *Teh*? The objective of this activity is to perform separation and purification of different types of tea and milk tea beverages.

Since milk tea is primarily a concoction of tea, with the addition of milk and sugar, we can make use of apparatus and simple set-ups to separate the various components accordingly. Two components that can be separated out are milk and sugar.

What can be used to extract milk and sugar from milk tea? An appropriate organic solvent can be used to extract the organic compounds in milk. Two different types of solvent (non-polar or polar) may be used. Sugar may be extracted via crystallisation.

Part A — Precipitation of Casein by the Addition of an Appropriate Organic Solvent to Form Calcium Caseinate

Precipitation, is a process where a precipitate is formed after two or more substances are added together. In this process, we will be able to separate precipitated Casein from milk.

Safety

Put on goggles, gloves, and lab coat. The organic solvent, hexane, should be handled in a fumehood.

Procedure

1. Pour 50 mL of *Teh-C* (see Figure 2) into a beaker.
2. Measure an equal volume of hexane and add into the same beaker. Two immiscible layers should form. Observe that the organic layer is in the top layer.
3. Repeat Steps 1 and 2 using ethanol instead of hexane. Observe the separation.
4. Filter the mixture using a filter funnel.
5. Observe the filtrate and residue. Retain the filtrate for Parts B and C.
6. Repeat the entire procedure for *Teh-O* and *Teh* (see Figure 2).
7. Compare the physical appearance of the filtrate and residue extracted from the three beverages.

Part B — Separation of Tea Extract from Ethanol Using Fractional Distillation

Fractional distillation is the separation of a mixture into its component parts (with different boiling points) by heating them to a temperature at which one or more fractions of the compound vapourise. This type of distillation is used when the boiling points of the components in the mixture have boiling points that differ by less than 25°C. Otherwise, a

Figure 2. A picture of *Teh-O*, *Teh-C*, and *Teh* (from left to right).

simple distillation could be used instead. In this case, the boiling point of ethanol is 78°C and the boiling point of water is 100°C. The small difference in boiling points requires the use of fractional distillation to separate the ethanol from the tea extract.

Procedure

1. Set up the experiment as shown in Figure 3, with Solution A collected from Part A in the round bottom flask. Use a hot water bath instead of a Bunsen burner.
2. Heat the mixture until the first drop of distillate (solvent) is collected at conical flask. Record this temperature, as this is the boiling point of the distillate.

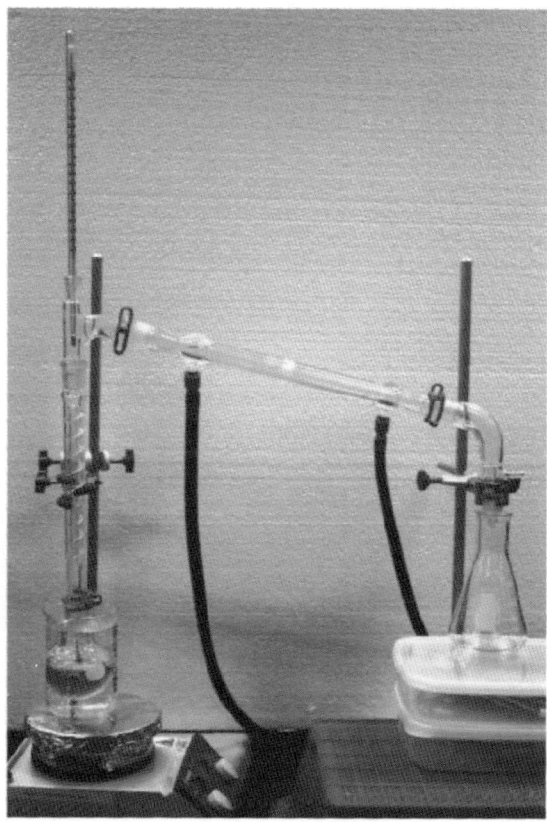

Figure 3.   Fractional distillation setup.

3. Continue to collect the distillate until the temperature rises or the distillate ceases to flow out of the condenser.
4. The liquid that is left behind in the round bottom flask contains the organic content from the milk tea.

Part C — Separation of Sugar from Aquatic Layer Using Crystallisation

Crystallisation refers to the formation of solid crystals from a homogeneous solution. It is essentially a solid–liquid separation technique, in which there is a mass transfer of a solute from the liquid solution to a pure solid crystalline phase occurs.

Procedure

1. Set up the tripod stand with a piece of wire gauze on top of the Bunsen burner. Light up a non-luminous Bunsen flame.
2. Place the beaker that contains the aqueous layer (Solution B) collected from Phase 1 on top of the wire gauze and heat the solution till it starts to boil.
3. Let the solution boil till its new volume is half its original volume. Stop heating and allow the saturated solution to cool down to room temperature.
4. Do not agitate the solution while it cools down.
5. Once the solution has cooled to room temperature, place the beaker carefully into an ice water bath to cool it further and allow crystallisation to occur.
6. Let the set-up sit until all the crystals are formed.

Inquiry Questions

Initial examination of milk tea

- What are the ingredients found in milk tea?
- What are the nutritional components that can be found in milk tea?
- Which of the components determine the health benefits or hazards of the beverage?
- How can we separate the various components for further examination?

Part A: Precipitation

- Why do the organic and aqueous layers not mix?
- Why is the organic layer below the aqueous layer?
- What is casein?
- Suggest a reason why it is preferable to use the 45 mL of solvent in three portions of 15 mL instead of one 45-mL portion for solvent extraction?
- Do you think it is a fair test to compare the nutritional values of the drinks in this aspect?

Parts B & C: Purification and crystallisation

- Can we evaporate the mixture solution to dryness to obtain sugar? Why?
- Why must we cool down the hot solution for crystallisation?
- What will happen if we agitate the solution during crystallisation? How does it affect the formation of crystals?
- What are the factors that affect the rate of crystallisation?

Chapter 6

Coconut Water

Lim Shan Yan and Boo Michelle

Background

After delivering some goods, Uncle Tan enjoys taking a breather at the kopi-tiam (traditional coffee shop; "kopi" is a Malay word for coffee and "tiam" means shop in the Hokkien or Hakka dialect) near his provision shop. He gave the drink stall lady a nod and she came by swiftly with a Thai coconut. As Uncle Tan took a long sip of the clear liquid from the fruit, a pleasantly refreshing sensation swept over him like a wave. A thought, *"Always the perfect thirst quencher on such a blindingly hot day"*, came to his mind. At the same time, he caught sight of the myriad of *nyonya* cakes at a stall run by a Peranakan lady. Other than the multicoloured *kueh lapis*, there were also *kueh dadar, kueh talam, kueh kochi* and *kueh kosui*.[1] The thought of his

---

[1] ***Kueh lapis*** or layered cake: A cake with thin, distinct alternating layers, usually dark and light brown. Its main ingredients are butter, eggs, and sugar.

Kueh dadar or omelette cake: A crepe coloured green due to flavouring with pandan juice extracted from pandan leaves. It is rolled into a cylinder, wrapping grated coconut, trickled with *gula Melaka*, a Malaysian palm sugar within.

Kueh talam or tray cake: A two-layered green-and-white cake. The sweet bottom green layer is made of green pea flour infused with pandan leaf extract while the white layer on top, which is slightly salty, consists of rice flour steeped in coconut milk.

adorable granddaughter came to Uncle Tan's mind and he smiled, "*I shall buy some kueh back for Little Annie, she loves them*".

The coconut is a popular fruit in Singapore and other Southeast Asian countries. The plant from which the fruit is grown, the *Cocus nucifera* palm, offers an extensive variety of uses, from its leaves, to the fruit and its husk, flesh, and water.

The drink that Uncle Tan enjoys is the coconut water. This is not to be confused with coconut milk, one of the ingredients in the *nyonya kueh*. Coconut milk is a white liquid extracted from grated coconut meat. Unlike coconut milk, coconut water refers to the clear liquid endosperm that is found in coconuts. Other than being served as a refreshing beverage, coconut water is known to offer many beneficial uses due to its properties. For instance, it is known to be sterile; this renders the liquid being used as an intravenous drip in the past as an alternative to saline solution (Petroianu *et al.*, 2004) and patients did not exhibit significant sensitivity towards it (Eiseman *et al.*, 1954). This has also been enacted in a scene in Jackie Chan's 1998 blockbuster "Who am I?" Coconut water has also been claimed to be a substitute for sugary sports drinks (Yong *et al.*, 2009). Due to the many benefits that coconut water offer, it is sometimes hailed as "the fluid of life" (Cocotap, 2013).

Objectives

This science activity aims to investigate: (1) the difference in the sugar content between young and old coconuts using Benedict's Test, and (2) the potential of coconut water as an electrolyte as it contains soluble minerals such as potassium.

Kueh kochi or Passover cake: A glutinous rice dumpling flavoured with pandan extract before being folded into a pyramid with a banana leave encasing a caramelised grated coconut filling.

Kueh kosui: A small rice cake shaped like a teacup, saturated with gula Melaka and pandan juice. It is best eaten with freshly grated coconut. This springy cake is then rolled in grated coconut before serving.

Description of Activities

Activity 1: The sugar content of young and old coconuts

Figure 1 shows the images of a young coconut and an old coconut. A typical young coconut has a green shell and a thick gelatinous layer of meat on the inside. On the other hand, a mature coconut appears brown as it has lost its green outer husk and its meat is much firmer. Young coconuts also contain more coconut water than old coconuts. It is often said that young coconut water is sweeter than old coconut water and there are two possible reasons to that. First, young coconut water might have a higher level of reducing sugars than old coconut water (Jasa, 2012). Second, young coconut water may have more fructose which is sweeter than sucrose (Tazzini, 2012).

The aim of this activity is to qualitatively compare the amount of reducing sugar in the coconut water of young and old coconuts. Benedict's test for reducing sugars is a semi-quantitative test which shows the relative amount of reducing sugars present. Reducing sugars such as glucose, fructose and maltose reduce the blue Benedict's solution containing copper(II) ions to insoluble reddish brown copper(I) oxide precipitate. Table 1 shows the different colours of precipitate and the relative amount of reducing sugars in the solution.

Figure 1.   Young (left) and old (right) coconut.

Table 1.   Colours of the precipitate in different amounts of reducing sugars.

Observation	Amount of reducing sugars
Blue solution	Reducing sugars are absent.
Green precipitate	Trace amount of reducing sugars is present.
Orange precipitate	Moderate amount of reducing sugars is present.
Brick-red precipitate	Large amount of reducing sugars is present.

The materials, reagents, and apparatus needed for this activity are listed below:

Materials/Reagents/Apparatus:

Young coconut
Old coconut
Benedict's solution
Distilled water
Bunsen burner
Beakers
Test tubes
White tile

Method

1. Add 2 cm^3 of coconut water into two separate test tubes.
2. Label the test tube containing young coconut water as Tube 1 and the tube containing old coconut water as Tube 2 (Figure 2).
3. Add 2 cm^3 of Benedict's solution and shake gently (Figure 3).
4. Heat the solution by immersing the test tubes into a boiling water bath for 5 minutes.
5. Observe the colour of the precipitate formed, if any.

Results

Figure 4 shows the coloured precipitate formed during the five-minute incubation period in the boiling water bath. After five minutes, Benedict's

Figure 2. Beaker 1 contains the coconut water from a young coconut. Beaker 2 contains coconut water from an old coconut. Beaker 3 contains distilled water and acts as a control.

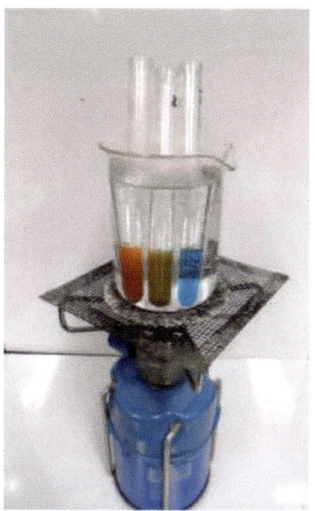

Figure 3. Benedict's solution was added to separate test tubes containing young coconut water, old coconut water, and distilled water (left to right). Precipitates form after heating the mixture for some time in a water bath.

solution was added to each test tube. The young coconut water formed a brick-red precipitate, while the old coconut water formed a brown precipitate. The distilled water served as a control to demonstrate the absence of reducing sugars where the blue Benedict's solution remained blue.

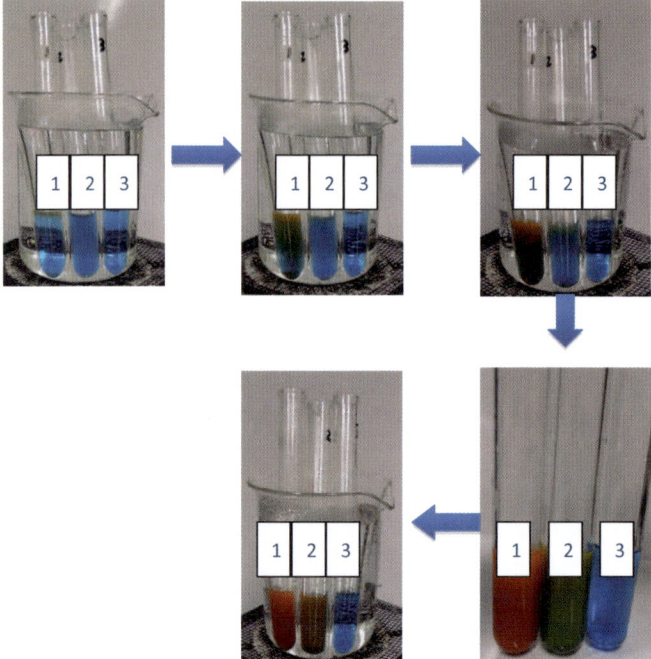

Figure 4.   The formation of precipitate during the Benedict's test.

Based on the intensity of the colour of the precipitate, we could conclude that both the young and old coconut water contained a large amount of reducing sugars.

The brick-red precipitate appeared earlier in the young coconut water than the old coconut water.

Discussion

The composition of the sugars in the coconut water varied throughout the ripening process. According to a previous study by Santoso *et al.* (1996), young coconut water contains a higher proportion of glucose and fructose. On the other hand, old coconut water contains a higher proportion of sucrose (Child and Nathanel, 1950). This explains why the brick-red precipitate was formed faster in the young coconut water than the old coconut water.

Sucrose is a non-reducing sugar that could be hydrolysed to form glucose and fructose. Our results showed that old coconut water formed a brown precipitate with Benedict's solution, which could possibly be due to the breakdown of sucrose into glucose and fructose in the old coconut water. Since old coconut water contains a higher proportion of sucrose, the delayed appearance of the brown precipitate could be due to the longer time needed for hydrolysis to occur.

Key

1. Young coconut
2. Old coconut
3. Distilled water

Post-lab activity questions

1. Comment on the effectiveness of the Benedict's test as a semi-quantitative test for reducing sugar.
2. The solutions were heated in the boiling water bath instead of being heated directly over the flame. Explain why.
3. How could the experiment be improved? List the reagents and apparatus required.

Inquiry questions

1. How can we apply the knowledge of varied sugar contents in young and old coconut water in our daily lives?
2. How can this Benedict's test be extended to other types of food substances for a meaningful analysis?

Activity 2: Coconut water as electrolytes

Can coconut water act as an electrolyte? This question is intriguing because it could show the possibility of using coconut water as a safe, natural, and alternative form of oral electrolyte (Munir, 1980) In rural communities, there are a significant number of people suffering from waterborne diseases and diarrhoea is one common symptom of such diseases. Other problems include microbial infections such as cholera and Hepatitis A. In these areas, commercial electrolytes in hospitals may not be adequate or accessible to every patient. Coconut water could be a suitable option as an electrolyte to rehydrate the patients.

To investigate the electrical conductivity of coconut water, electricity can be passed through coconut water. Observe whether a light bulb will light up. The quantity of charge that passes through the electrolyte is then measured. The materials and apparatus used in the investigation are given below. When graphite electrodes are dipped in the coconut water, the voltmeter and ammeter would register a change in voltage and current.

The materials, reagents, and apparatus needed for this activity are listed below:

Materials/Reagents/Apparatus:

Young coconut
Old coconut
Deionised water
Beakers
Ammeter
Voltmeter
Light bulb
Batteries
Wires with crocodile clips
Graphite electrodes

Method

1. Place 200 cm^3 of coconut water in two clean 250 cm^3 beakers.
2. Label the beaker containing young coconut water as Beaker (1) and the beaker containing old coconut water as Beaker (2).
3. Attach a graphite electrode to each end of a piece of wire (Figure 5). The wires should be connected to battery cells. Secure using crocodile clips.
4. Dip the two electrodes into the sample at the same time and at the same depth.
5. Record your observations on the ammeter/voltmeter (Figure 6).

Figure 5.   Graphite electrodes dipped in coconut water.

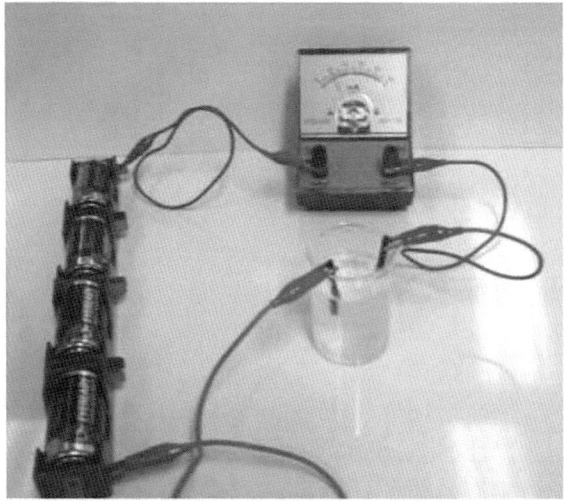

Figure 6.   Experimental set-up to measure the current across the coconut water.

Table 2.   Current and voltage registered when the electrodes are placed in each of the three samples shown below.

Sample	Current/mA	Voltage/V
Young coconut water	28	3.8
Old coconut water	34	4.0
Deionised water	0	0

Results

Table 2 shows the voltage and current readings when the electrodes were dipped in the young coconut water, old coconut water and deionised water respectively. Both the young and the old coconut water registered comparable amounts of current and voltage. The deionised water served as a control — zero current and voltage were registered in the absence of electrolytes.

Inquiry questions

How could the coconut water be processed to boost the concentration of the electrolytes?

What are some other practical uses of coconut water?

Explore Further! (Extension Activity)

Test for non-reducing sugars

Further tests can be performed to determine the amount of non-reducing sugars in the coconut water. By coupling this method with the tests for reducing sugars (e.g., Benedict's test), we can obtain more information on the types of sugar present in coconut water.

Try One of the Following Experiments!

Experiment 1

Additional materials you will need:

- Dilute hydrochloric acid
- Sodium hydrogen carbonate

1) Place 5 cm³ of coconut water in a clean test tube.
2) Add 2 cm³ of dilute hydrochloric acid.
3) Heat the solution by immersing the test tube into a boiling water bath for 5 minutes.
4) Neutralise the solution by adding 10% sodium hydrogen carbonate solution, checking using an appropriate pH indicator.
5) Perform the Benedict's test on the resulting solution.

* The hydrochloric acid hydrolyses the glycosidic bonds in the sucrose molecules. This separates and releases the glucose and fructose monomers (Sadasivam and Manickam, 1996).

Experiment 2

Additional materials you will need:

• Invertase or sucrase

1) Place 2 cm^3 of coconut water in a clean test tube.
2) Add 0.2 cm^3 of invertase or sucrase.
3) Incubate the mixture under room temperature for 30 minutes.
4) Perform the Benedict's test on the resulting solution.

* Invertase, or beta-fructofuranosidase, is an enzyme that catalyses the breakdown of sucrose. It is derived from yeast and is used primarily in confectioneries to produce liquid centers in candies (Systems, 2009).

What does it tell us about the sugar content of coconut water if the Benedict's test gives a brick-red precipitate in this experiment?

Ms. Coco$_3$

Answer

If there are non-reducing sugars present in the hydrolysed sample, we would observe a heavier precipitate formed in the Benedict's test, as compared to the sample before hydrolysis.

How Sweet is Sweet?

"What's in a name? That which we call a rose. By any other name would smell as sweet"

—William Shakespeare

Is the Water from a Young Coconut Sweeter Than That from an Old Coconut? How Do You Measure "Sweetness"?

Fructose is said to be much sweeter than sucrose (Tazzini, 2012). And there are claims that young coconut water tends to have a sweeter taste compared to old coconut water. But is it always the case? Could any of the following factors affect this?

- Temperature?
- pH?
- Physical properties of the sugars?

References

Child, R., Nathanel, W. R. M. (1950). Changes in the Sugar Composition of Coconut Water during Maturation and Germination. *Journal of the Science of Food and Agriculture*, 1(11), 326–329.

Cocotap. (2013). Understand the Benefits of Coconut Juice. Why Does It Feel So Good? Available at http://www.cocotap.com/nutrition.htm.

Eiseman B, Lozano, R. E., Hager, T. (1954). Clinical Experience in Intravenous Administration of Coconut Water. *A.M.A. Archives of Surgery*, 69(1), 87–93.

Jasa. (2012). Composition of Young Coconut Water and Old Coconut Water. Available at http://jasareviewmurah.wordpress.com/2012/02/04/composition-of-young-coconut-water-old-coconut/.

Munir, M. M. I. (1980). Coconut Water as One of the Optional Oral Electrolyte Solutions. *Paediatrica Indonesiana*, 20(1–2), 38–46.

Petroianu, G. A., Kosanovic, M., Shehatta, I. S., Mahgoub, B., Saleh, A., Maleck, W. H. (2004). Green Coconut Water for Intravenous Use: Trace and Minor Element Content. *The Journal of Trace Elements in Experimental Medicine*, 17(4), 273–282.

Sadasivam, S., Manickam, A. (1996). *Biochemical Methods*. New Age International Publishers.

Santoso, U., Kubo, K., Ota, T., Tadokoro, T., Maekawa, A. (1996). Nutrient Composition of Kopyor Coconuts (*Cocos nucifera L.*). *Food Chemistry*, 57(2), 299–304.

Systems, G. H. (2009). Invertase. Available at http://greenwoodhealth.net/np/invertase.htm.

Tazzini, N. (2012). Invert sugar. Available at http://www.tuscany-diet.net/2012/06/08/invert-sugar/.

Yong, J. W., Ge, L., Ng, Y. F., Tan, S. N. (2009). The Chemical Composition and Biological Properties of Coconut (*Cocos nucifera L.*) Water. *Molecules*, 14(12), 5144–5164.

Chapter 7

No Durian On MRT!

Low Wei Chuan Matthias

Food is one of the most prominent cultural elements that characterise a culture. People of the same culture would invariably have similar food types. Consequently, the manner in which food is acquired and consumed reflects the local culture and way of life. The important link between food and culture can also be seen in many parts of the world. For example, rice is a common staple food and tea a common beverage in the Chinese community.

Food has been such an integral part of the Singapore society. Very often, conversations amongst Singaporeans revolve around food. The love for food is clearly evident even on social media platforms such as the Instagram where people upload photos of the food they eat. Therefore, using localised food items and their characteristics to develop scientific concepts could provide an authentic learning experience for students.

Introduction

King of fruits

Although the flesh is potently odoriferous, durian is as much a cultural icon; it is a treasured and popular fruit in Southeast Asia. Widely known as the "King of fruits" because of its tough thorn-covered husk, durian is a popular fruit amongst Singaporeans and Southeast Asians. The fruit is native to Southeast Asian countries such as Indonesia, Malaysia, and

Figure 1.   Durian sold at a fruit stall. Durian is typically wrapped and sold in Styrofoam boxes to keep out its strong smell.

Thailand due to the ambient weather. During the peak production season, durians are imported and sold on the streets of Singapore. Besides its spiky appearance, the fruit has a strong and unique smell that characterises the fruit. The famous French explorer Henri Mouhot commented, "On first tasting it [durian] I thought it like the flesh of some animal in a state of putrefaction." The pungent smell of durian can be offensive and overpowering to those who do not have an acquired taste for it. Particularly in Singapore, durians are not allowed in public transport due to its unpleasant smell. Yet, durian evokes an entirely different response from durian lovers: they would find its flesh rich and creamy.

The Process of Diffusion

The process of diffusion has many important applications spanning across different areas in the three branches of science: chemistry, biology, and physics. From the exchange of gases during respiration and photosynthesis in both plants and animals to pollination to explaining Brownian motion in thermal physics, the process provides evidence for the particulate nature of matter and explains key biological processes in living organisms. Hence, the topic on diffusion is of great importance in the school science curriculum.

However, a number of research have highlighted that it is difficult to bring across the concept in a meaningful manner to allow students to understand the importance of the process of diffusion (Christianson and Fisher, 1999). In addition, learning difficulties on the concept of diffusion were also identified in various studies over the years (Tekkaya, 2003; Odom, 1995). For instance, none of the respondents in a research study attributed dye diffusion in water to the random motion of particles (Westbrook, 1991).

Objectives

As students come to school with a host of different personal experiences and ideas about diffusion, the activity described in this chapter seeks to present the process of diffusion in a culturally relevant manner using durian as the cultural item to connect with daily phenomenon that students encounter. In this way, students will be able to connect their preconceptions with scientific knowledge that they learn in the classroom to help them overcome any misconception.

The objective of this activity is to enable students to explain everyday phenomenon of diffusion using concepts such as kinetic particle theory and convection current. Also, students should be able to explain the effect of temperature on the rate of diffusion. It is also intended for students to develop critical thinking skills through the application of scientific knowledge to explain daily occurrences that they see or experience in their everyday lives.

Activity

Work collaboratively in a group to explain a given scenario below:

> *In Singapore, durians are not allowed to be brought into a Mass Rapid Transit (MRT) train due to its pervasive pungent smell. Prominent signage like the ones below can be seen on public transport systems:*
>
> *John, a foreigner, was fined by the train authority for bringing durian into the train cabin one day. He does not know the basis of the restriction.*
>
> *As the representative of the train authority, you are required to explain to John the basis of the restriction based on scientific principles. In your group, craft a written response of 100–150 words to explain to John the basis of the restriction based on scientific principles.*

Inquiry Questions for Exploration

- Where do durians originate from? What is the scientific name of durian?
- What are some characteristics of durian? State one distinct feature of durian compared to other tropical fruits?
- What are some food items made of durian?
- What are some uses of durian in daily life? What are some health benefits of consuming durian?
- What are the soil and climatic requirements (pH and temperature etc.) for durian to grow well?
- Why are you able to detect the presence of durian from afar? Will you be able to detect the smell of durian in a vacuum room?
- How does the rate of diffusion relate to molar mass of a gas? Find this out using Graham's Law of Diffusion.
- What is/are the chemical compound(s) responsible for the pungent smell of durian? How do you think these substances can be analysed?
- Durian husks are a major source of waste. How do you think it can be recycled? What are some of its characteristics that enable it to be recycled?
- Durian is known to contain tryptophan, an amino acid. How is tryptophan related to the serotonin, a monoamine neurotransmitter? What are some functions of serotonin?

- Some believed that durian is "heaty" while durian husks are "cooling". What is meant by "heaty" and "cooling" in this context?

References

Christianson, R. G., Fisher, K. M. (1999). Comparison of Student Learning About Diffusion and Osmosis in Constructivist and Traditional Classroom. *International Journal of Science Education*, 21(6), 687–698.

Odom, A. L. (1995). The Development and Application of a Two-tiered Diagnostic Test Measuring College Biology Students' Understanding of Diffusion and Osmosis Following a Course of Instruction. *Journal of Research in Science Teaching,* 32, 45–61.

Tekkaya, C. (2003). Remediating High School Students' Misconceptions Concerning Diffusion and Osmosis through Concept Mapping and Conceptual Change Text. *Research in Science & Technological Education*, 21, 5–16.

Westbrook, S. A. (1991). A Cross-age Study of Student Understanding of the Concept of Diffusion. *Journal of Research in Science Teaching,* 28, 649–660.

Additional resources:

Burhill, I. H. (1966). *A Dictionary of the Economic Products of the Malay Peninsula,* pp. 885–889. Kuala Lumpur: Ministry of Agriculture and Co-operatives.

Hutton, W. (1996). *Tropical fruits of Malaysia & Singapore,* pp. 20–21. Hong Kong: Periplus Editions.

Piper, J. M. (1989). *Fruits of South East Asia: Facts and Folklore*, pp. 17–22. Singapore: Oxford University Press.

Chapter 8

Sodium Sulphite in Chicken Frank

Khoh Rong Lun

In 2012 alone, Singapore imported a total of 183,343 tonnes of chicken (http://www.ava.gov.sg/Publications/Statistics/), and in the financial year 2012/2013, 57 new establishments were accredited by Agri-Food & Veterinary (AVA) for import of process chicken, amongst other meat. (2012/2013 Corporate Annual Report, AVA, http://www.ava.gov.sg/NR/rdonlyres/0676D1EB-C401-4038-9D8D-84A01B52DD27/26874/AVA_AR2013Lores.pdf). There is a sustained appetite for processed meat due to its long shelf life and convenience in preparation that matches the needs of busy working adults in Singapore. It is also often a choice food at events like barbeque (BBQ) get-togethers which is a popular leisure activity.

Chicken frank is one popular form of processed meat. Besides offering convenience in preparation, it is typically added as additional ingredient for one-dish meals such as Nasi Lemak, fried rice, and instant noodles. Chicken with cheese fillings can also be found at the *pasar malam* (local make-shift night markets). Another contributing factor to it being a food choice is its price affordability; it can cost lower than 60 cents per 100 grams (Donato *et al.*, 2008).

Chicken frank may contain sodium sulfite (sulphites) (E221) (Figure 1) as a preservative (Figure 1). Although sodium sulphite has limited use in the actual preservation of meat, it is often added for its ability to maintain

(A)

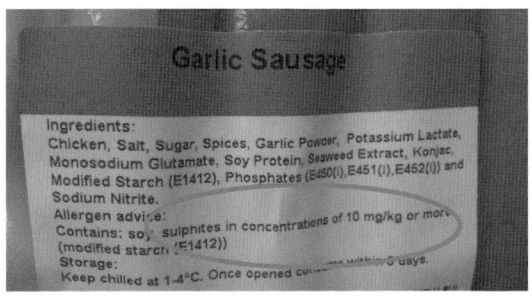

(B)

Figure 1. (A) Cheese sausage and (B) garlic sausage found in supermarkets. Both reportedly contain sulphite, as reflected under the allergen advice section of their labels.

brilliant red colour in meat even upon exposure to air giving the perception that the meat is fresh.[1]

However, it is unknown to many that sodium sulphite is a form of potential trigger for breathing difficulty and may reduce vitamin content in products (Harrington, 1904).[2] Thus, its quantity in processed food ought to be carefully monitored. The recommended daily intake is at no more than 0.7 mg kg^{-1} body weight (Food Science Australia, 2006).

---

[1] http://www.food-info.net/uk/e/e221.htm.
[2] http://www.articlesbase.com/home-brewing-articles/food-sulphites-the-asthma-culprit-5610054.html.

Objective of the Activity

Owing to Singapore's inadequate farming and agricultural work, she is highly reliant on the food imported from other countries. Coupled with the projected increasing overall population figures, and hence greater food consumption, the gravity of food safety issue cannot be neglected. While Singapore has always been known for her high standards towards quality control and food safety, she is not impervious to episodes of food scares. One of the incidents in our recent collective memories would be having traces of maleic acid found in tapioca starch-based *pearls* used in *bubble tea* which is believed to contribute to the chewy texture of the said food as reported in 2013.

This activity presents an opportunity for the discussion of national issues of reliance on resources from other countries as well as importance of food safety. Concepts of redox chemistry and titrimetric analysis will be applied in the analysis of the amount of sodium sulphites found in different brands of chicken franks found in Singapore.

Description of the Activity

The main processes involved can be divided into three parts: (i) pre-treatment (ii) extraction of sulphite content and (iii) titrimetric analysis.

Pre-treatment

Sample pre-treatment involves converting the chicken frank into a form that is more easily analysed. To ensure that the sulphite content can be more accurately determined, the surface area-to-volume ratio of the sample should be maximised. This can be done by blending the chicken frank.

Apparatus needed:
    Food blender
    Mass balance
    250 cm³ glass beaker
    200 cm³ three-neck round bottom flask

Figure 2.   Blended chicken frank.

Procedure:

1. Weigh about 500 g of chicken frank.
2. Transfer the chicken frank into a food blender. Blend the chicken frank until it is visibly mashed up.
3. Transfer the blended sample (see Figure 2) into a clean and pre-weighed dry 200 cm^3 3-neck round bottom flask. Weigh and record the exact mass of the blended sample.

Points to consider:

1. To what degree of accuracy should the mass readings be recorded to?
2. Is it suitable to use the tare function of the electronic mass balance in this instance?

Extraction of sulphite content

The extraction process of sulphite content follows the Monier–Williams method. The sulphite content in the sample would be converted to sulphur dioxide when boiled with hydrochloric acid. Subsequently, the sulphur dioxide gas would be swept by a stream of nitrogen gas into hydrogen peroxide which acts as an absorbing and oxidising agent. Sulphur dioxide would be oxidised into sulphuric acid. The amount of sulphuric acid present can then be quantitatively determined by titrating using a standard solution of sodium hydroxide.

Set-up:

Apparatus needed:

1. 250 cm³ dropping funnel
2. 250 cm³ round-bottom flask
3. 150 cm³ conical flask
4. Liebig condenser
5. Rubber tubes with connectors
6. Retort stand

Procedure:

1. Set up the apparatus as shown in Figure 3.
2. Test the set-up by passing nitrogen gas through and observe for bubbling in the hydrogen peroxide solution.

Figure 3. Set-up for the extraction of sulphite from chicken frank sample.

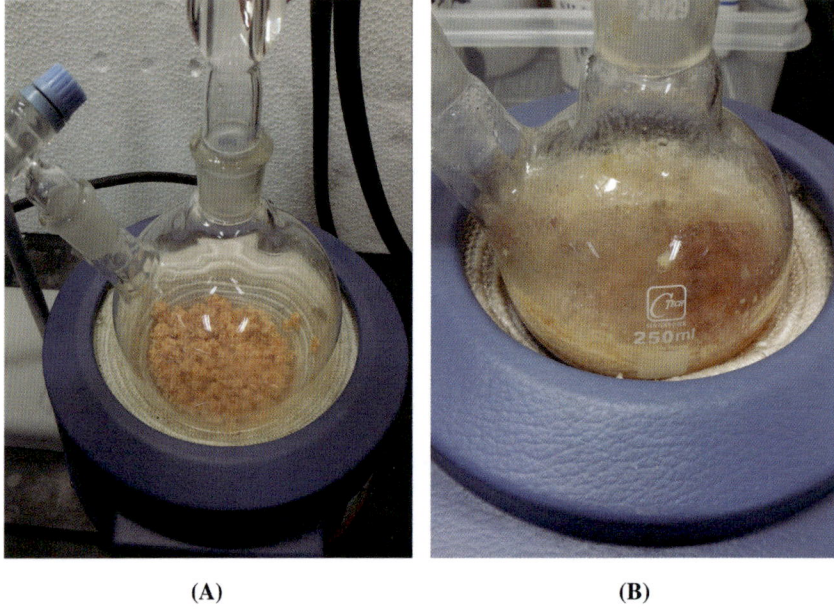

(A)   (B)

Figure 4.   Close-up of chicken frank in sample (A) before and (B) during reflux.

Figure 5.   Close-up of the conical flask containing hydrogen peroxide. Bubbles of SO_2 (as indicated by arrow) gas was purged into the hydrogen peroxide.

3.   Allow the nitrogen gas to flow for 15 minutes to have an inert atmosphere within the set-up.

4.   Add in the hydrochloric acid from the dropping funnel and start the reflux. Allow the reflux to be carried out for 105 minutes (Figures 4 and 5).

5. Switch off the power supply to the heating mantle while allowing the nitrogen gas to still bubble.
6. Add in 2–3 drops of phenolphthalein indicator to the hydrogen peroxide solution.

Points to consider:

1. Other than phenolphthalein, what other indicators may be used?
2. How do you decide which type of indicator is suitable for the titration?
3. Why is only a small volume of indicator used?

Titrimetric analysis

Using a standard solution of 0.0100 mol dm^{-3} sodium hydroxide, determine the amount of sulphur dioxide formed using titrimetric analysis.

Apparatus needed:

1. 50.00 cm^3 burette
2. Retort stand with clamp

Chemicals needed:
0.0100 mol dm^{-3} sodium hydroxide

Procedure:

1. Fill the 50.00 cm^3 burette with the 0.01 mol dm^3 sodium hydroxide.
2. Titrate until a permanent pink is observed in the hydrogen peroxide solution (Figure 6).
3. Repeat the entire experiment (starting from the pre-treatment) until at least two consistent titre values (within ±0.10 cm^3) are obtained.

Points to consider:

1. What is the percentage error associated with each titre value determined?
2. Why is a strong base like sodium hydroxide used? Can it be replaced with a weak base such as ammonia?

Figure 6.   Titration carried out until a permanent pale pink is observed.

Data analysis

Let the average titre value detemined be x cm^3.

Derive an expression, in terms of x, the concentration (in parts per million, ppm) of the sulphite content in the chicken frank.

Inquiry questions

1. What are some possible sources of error in this experiment?
2. What are some challenges in the use of Monier–Williams method? What alternatives are there in analysing sodium sulphite content in food? How can the experimental setup be simplified?
3. Due to the possible side effects of sodium sulphite, its use has been reduced. What other food additives had been used in place of sodium sulphite? What are some advantages and/or disadvantages of using the alternative additive?
4. Sodium sulphite is also known to be found in some beverages such as wine and fruit juices. How would the method of analysis change if the food sample is in the liquid form?

5. Why are "E" numbers used to denote food additives? What are some possible disadvantages of using "E" numbers instead of the actual names?

Notes for Users

The quantity of sulphite present is small and will most likely require multiple repeats of extraction of sulphites before a substantial amount can be collected for proper titrimetric analysis can be carried out.

References and Additional Recommended Resources

Agri-Food & Veterinary, Singapore (2013). Press release: Recall of Starch-Based Products From Taiwan Due to Maleic acid.

Donato, R. M., Toledo, M. C. F., Almeida, C. A. S., Vicente, E. (2008). Analytical Determination of Sulphites in Fruit Juices Available on the Brazilian Market. *Brazilian Journal of Food Technology*, 11(3), 226–233.

Food Science Australia (2006). Meat Technology — Information Sheet (Sulphur dioxide, Sulphites in Meat Products).

Harrington, C. (1904). Sodium Sulphites: A Dangerous Food-Preservative. *The Journal of Infectious Diseases*, 1(2), 355–357.

Chapter 9

Biodegradable Tableware: Conserving Our Environment Through the Use of Materials from Renewable Resources

Tan Yong Leng Kelvin

Disposal Polystyrene Tableware: Impact on Our Environment

The next time you are at a party or having a takeaway meal, take a closer look at the disposable tableware (cutlery, plates, bowls, and cups) (Figure 1) that you are using: what material is it made of? Most likely, your disposable tableware will be made from the synthetic polymer polystyrene (see Figure 2). Plastics manufactured from petrochemicals, such as polystyrene, have become indispensable in our lives with their numerous uses and applications. Unfortunately, one of the key advantages that plastics offer us — that of durability — has generated global concern due to the incompatibility of these plastics with the environment after disposal. There is evidence to show that plastics are likely to last for hundreds of years before the material is broken down. Litter from plastics waste is ecologically undesirable and detrimental to marine life; incineration contributes to air pollution and may also generate toxic gases, while suitable landfill sites are limited.

Figure 1.   Biodegradable tableware made from a composite material combining corn-starch and polystyrene.

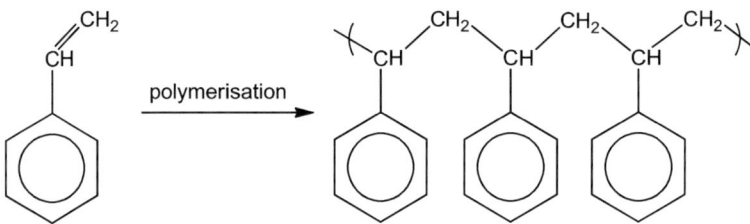

Figure 2.   Three repeating units of the chemical structure of polystyrene is shown on the right. Polystyrene, a synthetic polymer made from the monomer styrene, is one of the most widely used plastics.

Despite the potential detrimental consequence on our environment, there is still widespread use of disposable polystyrene tableware in Singapore. Takeaways from coffee shops, restaurants, and hawker centres, as well as catered food at events, mainly use polystyrene, tableware for reasons of convenience or as a means of cutting down on labour cost and water usage. The polystyrene tableware is disposed off as refuse after usage, and contributes to the waste problem in Singapore. In 2012, an average of 1370 kg of waste was generated per person in Singapore. 40%

of the waste was disposed with plastics making up 25% of the total waste disposal. Only 10% of the plastics waste was recycled.[1] The recycling symbols for different polymer types of plastics are given in Figure 3. These symbols, found on many plastic products, allows for the efficient identification and separation of waste plastic for recycling and reprocessing the materials into useful products.

Can we reuse or recycle our disposable tableware? Reusing tableware made from polystyrene is not encouraged as they may degrade with repeated washing and use. The design of the polystyrene tableware results

Figure 3. Recycling symbols for plastics.[2] The arrows cycling clockwise to form rounded triangles indicate generic recyclable material, while the number and acronym represent the different types of plastic polymers. For example, polystyrene is represented by the number 6 and the acronym PS.

---

[1] The 2013 statistics are available from NEA's website: http://app2.nea.gov.sg/energy-waste/waste-management/waste-statistics-and-overall-recycling.

[2] *Source of the image*: http://www.quantumbalancing.com/recycle.htm. Permission was granted to use this picture.

Figure 4.   The Semakau landfill is created by the amalgamation of two offshore islands Pulau Semakau and Pulau Sakeng. Designed to be in harmony with the surrounding eco-system, the flourishing plant and wildlife attract nature lovers who engage in recreational activities on Pulau Semakau's western end. Copyright © National Environment Agency of Singapore, 2013. All rights reserved.

in pockets of spaces that can harbour food particles and gives rise to rapid bacteria growth, even after repeated washing. On the other hand, polysty-rene is a thermoplastic and can be melted and remoulded repeatedly, so recycling is possible. However, it is presently not cost effective or viable to recycle used plastic tableware, as there is a need to separate them from other wastes, followed by washing to remove the food contaminants before they can be made into new products. In Singapore, used disposable table-ware is incinerated in an incineration plant. The ash obtained from refuse incineration, together with other non-recyclable, non-combustible parts of Singapore's waste, is buried in the country's sole landfill site, Semakau Landfill, located offshore in the southern part of Singapore (see Figure 4).

Alternatives to Polystyrene

The resistance of plastics such as polystyrene to chemical and biological degradation, coupled with a low recycling rate, has become a cause for concern and clearly there is a need to explore more sustainable and environmentally friendly alternatives. One option that offers potential in conserving both our environment and petroleum resources is the use of biodegradable plastics to replace conventional plastics. Biodegradable plastics are defined as plastics which can be degraded by the action of microorganisms such as bacteria, fungi, and algae, to give carbon dioxide, water, inorganic compounds and biomass. In the process of biodegradation, the polymer chain is first broken into shorter segments by extra-cellular enzymes via chemical reactions like oxidation and hydrolysis, followed by transportation of these fragments into the cell where they are biologically assimilated to give carbon dioxide, water, inorganic salts and biomass.

So far, biodegradable polymers have not completely replaced disposable polystyrene tableware. The best alternatives currently are composites that blend natural biodegradable polymers such as starch and cellulose with petroleum-based polymers like polystyrene. The presence of starch and cellulose, which are obtained from renewable resources such as plants, help enhance the biodegradation of the plastic. Although such composites will degrade after a certain amount of time if they are buried in landfills (or even in your own backyard), the biodegradation of these materials are optimised if carried out in a specially built composting facility, where conditions such as type of microbes, temperature, humidity, and lighting can be carefully adjusted to increase the rate of degradation. Due to the lack of suitable composting facilities in Singapore, used biodegradable tableware is incinerated to give humus, which can be utilised as fertiliser. Furthermore, the carbon neutrality function of biodegradable tableware ensures that there would be no net increase in carbon dioxide in the air from burning the portion made from biodegradable polymers. This is because the plants used to synthesise the biodegradable tableware initially absorbed carbon dioxide to form glucose, the building block of starch and cellulose.

Chemical Composition

Commercially available biodegradable tableware is made of composite materials that combine polystyrene with starch obtained from corn and yam. The presence of starch in biodegradable tableware can be confirmed by the addition of iodine solution, as shown in Figure 5. Starch is a mixture of two different polymers, amylose, and amylopectin, made up of

Figure 5.   Sample of biodegradable tableware before (left) and after (right) addition of iodine solution. The blue-black coloration indicates the presence of starch in the biodegradable tableware.

glucose monomer units linked together by glycosidic bonds (Figure 6). The glycosidic bonds can be hydrolysed under the influence of enzymes such as amylase which are present in human saliva and in the pancreas. These enzymes act as catalysts for the hydrolysis reaction under conditions compatible with human body temperature and approximately neutral pH. Alternatively, treating starch with acids or bases will also result in the glycosidic bonds being hydrolysed.

Guiding Questions

The hydrolysis of starch is an important part in the biodegradation process for biodegradable tableware. Although it may not be feasible to replicate the exact conditions used in the composting facilities, it is possible to carry out small-scale experiments to investigate the degradation of starch and the consequence on biodegradable tableware. It would be interesting to explore some of the conditions that would enable the glycosidic bonds in starch to be hydrolysed at a faster rate, and to investigate the effect of hydrolysis of the glycosidic bonds in starch on the physical properties of biodegradable tableware.

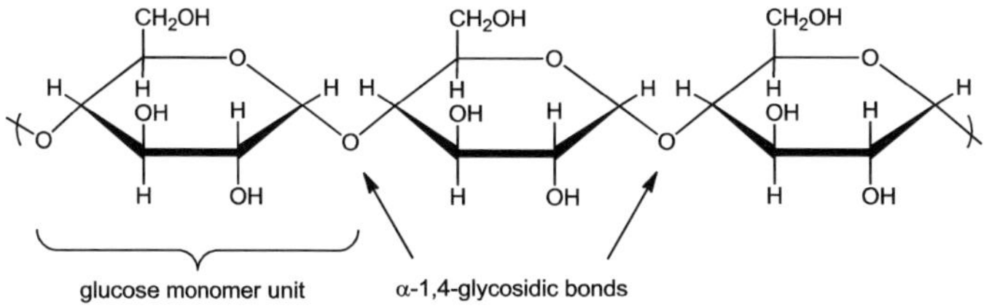

(A)

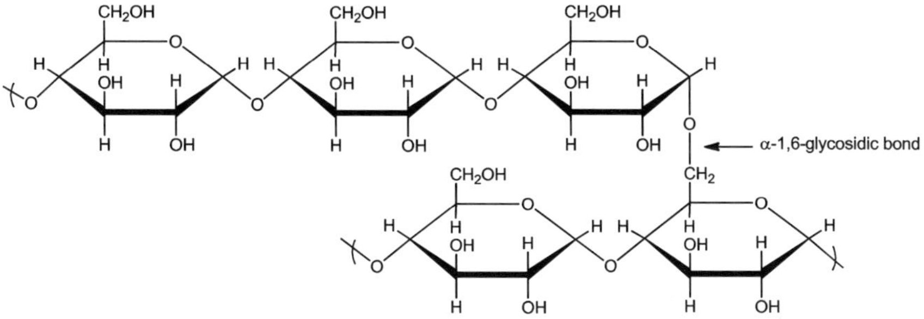

(B)

Figure 6. Amylose (**A**), the unbranched form of starch, consists of glucose monomer units linked via α-1,4-glycosidic bonds. Amylopectin (**B**), the branched form, has about one α-1,6-glycosidic bond per 24 to 30 α-1,4-linkages.

Effect of pH on Enzyme Catalysed Hydrolysis of Starch

Fill each of seven test tubes with 3 mL of 1% of starch solution and 3 mL of buffer solution (pH 1, 3, 5, 7, 9, 11, 13) respectively. Place the test tubes in a 37°C water bath. After five minutes, add five drops of pre-collected human saliva to each of the test tubes, and tap on the test tubes gently to ensure even mixing before returning the test tubes to the water bath. At regular time intervals of three minutes, place one drop of solution from each of the test tubes (using a different dropper for each solution) onto a petri dish and add one drop of iodine solution to each of the solution drops. The formation of a blue-black colour confirms the presence of starch; absence of the blue-black coloration indicates that the starch has been hydrolysed (Figure 7). The time required for the

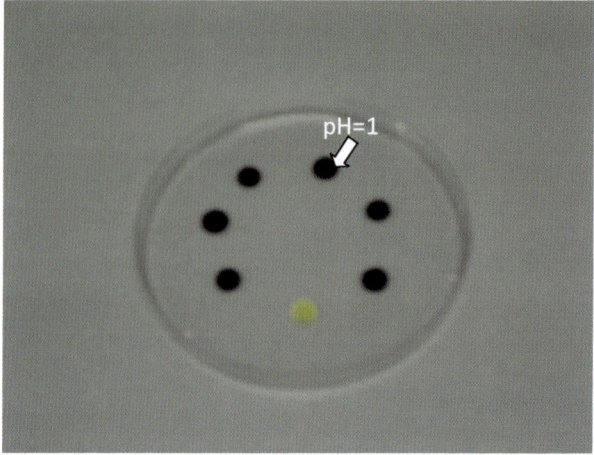

Figure 7.   Clockwise from top right: effect on hydrolysis of starch by human saliva in solutions of pH 1, 3, 5, 7, 9, 11 and 13 after 15 minutes (after addition of iodine solution).

hydrolysis of starch can be correlated to the relative enzyme activity in solutions of different pH.

Measurement of Tensile Strength

A simple device can be made to measure the tensile strength of the biodegradable tableware (Stevens and Poliks, 2003).[4] The tensile strength is the measurement of the stress (force per unit area) required to break the test specimen when it is placed between the two clamps and drawn.

Place samples of biodegradable tableware in buffer solutions of different pH values, as shown in Figure 8. Remove the samples from the solutions after two weeks and dry them. Using a template, cut out dog-boned shape specimens from the samples (such as the one shown in Figure 9) and clamp the specimens in place on the device. Subject the specimen to increasing stress either by pulling vertically downwards via a force sensor attached to a data logger (Figure 10), or by adding weights (Figure 11); the specimen will eventually break. By dividing the total load at the break point (units in newtons) by the original cross-sectional area (units in square meters) that broke, the tensile strength (units in megapascals) can

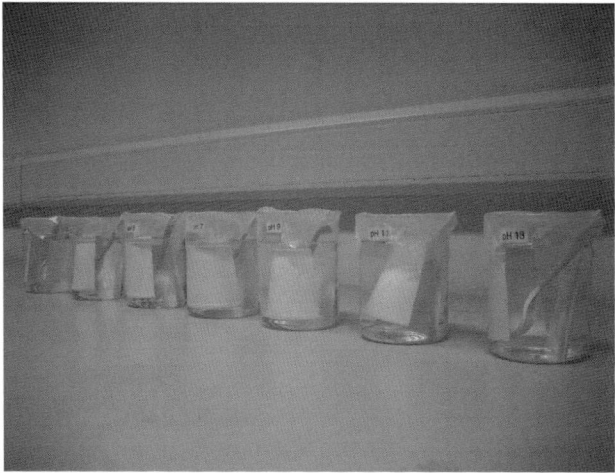

Figure 8.   Samples of biodegradable tableware placed in solutions of different pH.

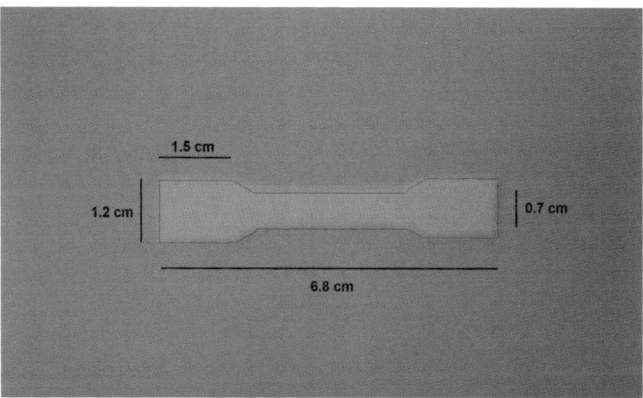

Figure 9.   Dog-boned shaped specimen cut out from a biodegradable plate.

be calculated. For example, if a specimen of original cross-sectional area of 0.64 cm × 0.0076 cm = 4.9 × 10⁻⁷ m² breaks with a load of 1.0 kg, the tensile strength (ts) in megapascals (1 MPa = 10⁶ N m⁻²) is:

$$ts = \frac{1.0 \text{ kg} \left(9.80 \frac{N}{kg}\right)}{4.9 \times 10^{-3} \text{ cm}^2 \left(10^{-4} \frac{m^2}{cm^2}\right)} = 2.0 \times 10^7 \text{ Pa} - 20 \text{ MPa}.$$

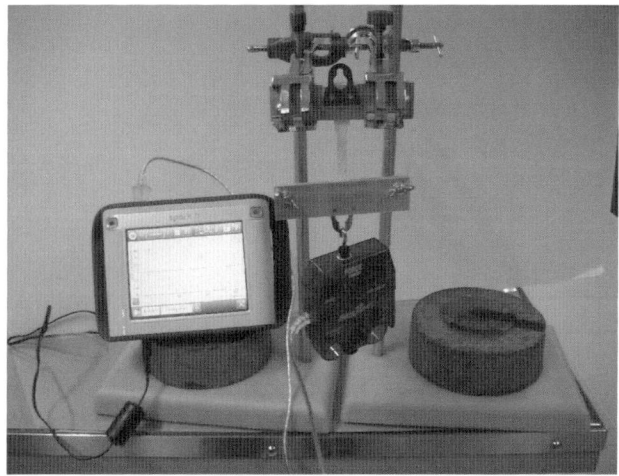

Figure 10.   Measuring the tensile strength using a force sensor and data logger.

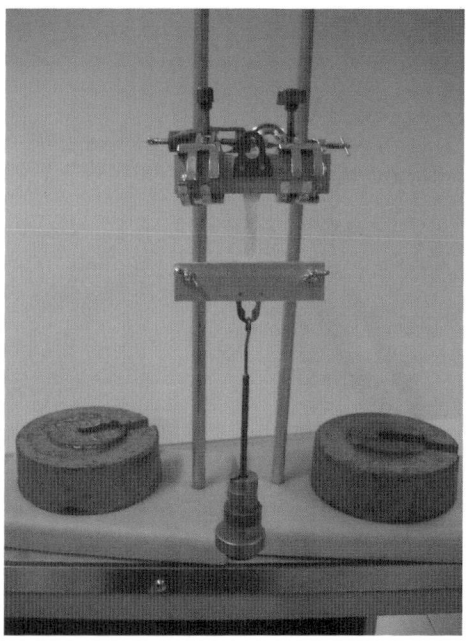

Figure 11.   Measuring the tensile strength using weights.

The tensile strength of the material before and after it has been placed in the buffer solutions can be determined and compared.

The Challenges Ahead

Presently, research into biodegradable plastics is focused on designing polymeric materials from renewable resources. These novel materials should have comparable physical and mechanical properties in terms of strength and stability to those of traditional non-biodegradable plastics. There is also the additional challenge of lowering manufacturing costs so that these biodegradable materials can be commercially produced for the mass market (Luckachan and Pillai, 2011). Until that goal is achieved, it is good to know that we can still do our individual part to contribute to the sustainability of our environment, by choosing to use biodegradable disposable tableware instead of the ones made from polystyrene the next time we organise a party or have a takeaway meal. An even better alternative would be to use non-disposable tableware which can be washed and reused.

Inquiry Questions for Students

1. Plastics can be categorised into thermoplastics and thermosets. What are the differences in the structures of thermoplastics and thermosets, and how do these structures contribute to their physical properties?
2. Polymerisation reactions can be classified as either addition or condensation reactions. What are the differences between these chemical reactions? Can you give examples of some known polymers which are formed by each of these types of reactions?
3. Recycled plastics can be made into a variety of products. What are some of the products that you know of that are made from recycled plastics?
4. In addition to biodegradation, what are some other means by which polymers can be made to degrade in a much shorter time compared to conventional petroleum-based plastics?
5. Biodegradable polymers have the potential for widespread use in the field of medicine and surgery. Can you suggest how biodegradable polymers would be useful in this field?

6. Bacteria are prokaryotes while fungi and algae are eukaryotes. What are some of the differences between prokaryotes and eukaryotes?
7. Can you describe some of the ways by which cells import and export their substances to produce nutrients and discard waste?

Author's Note

- In addition to varying the pH of the solutions for each of the two activities described above, explore other physical and chemical conditions that are likely to influence the rate of hydrolysis of starch as well as other means to measure the effect on biodegradable tableware.
- Many current scientific challenges require an interdisciplinary approach involving a variety of expertise. The topic presented here — that of biodegradable materials — may also be approached in a similar manner.

References

Luckachan, G. E., Pillai, C. K. S. (2011). Biodegradable Polymers — A Review on Recent Trends and Emerging Perspectives. *Journal of Polymers and the Environment*, 19, 637–676.

Stevens, E. S., Poliks, M. D. (2003). Tensile Strength Measurements on Biopolymer Films. *Journal of Chemical Education*, 80, 810–812.

Chapter 10

Colour Matters!

Lim Jia Ying Jessica, Cho Wen Jing and Gan Ghim Kui

Which school has the best uniform? This is a common discussion topic among students of different schools in Singapore, such that there are internet polls on "Which secondary school uniform is the coolest?" in a popular Singapore blogpost.

Since students' take much pride in their school uniform, it would be interesting for students to investigate the question: "What makes a school uniform '*cool*'?"

From the study of thermal physics, we understand that the colour and material of cloth affects the body temperature. Students can investigate the "coolness" of their school uniform as compared to other school uniforms.

The hot and wet climate of Singapore makes this experiment even more relevant for students to learn the science of heat absorption and heat emission so as to determine an apt clothing to wear at different weathers.

Objective of the Science Activity

We are familiar with the idea that white is the poorest absorber/emitter of heat while black is the best absorber/emitter. What about all the colours in between?

What makes a school uniform "cool"? What colour would be the most comfortable when choosing a T-shirt for outdoor and indoor activities? This activity will help us answer these questions.

Prior Knowledge

The higher the sensor readings, the better heat absorber or emitter the T-shirt will be.

Materials

1. Different coloured T-shirts of the same fabric (Figure 1)
2. Thermosensor
3. Multimeter
4. Stopwatch
5. Tunable heat source, e.g., tungsten lamp or hot plate
6. Digital camera (with adjustable exposure and shutter speed)

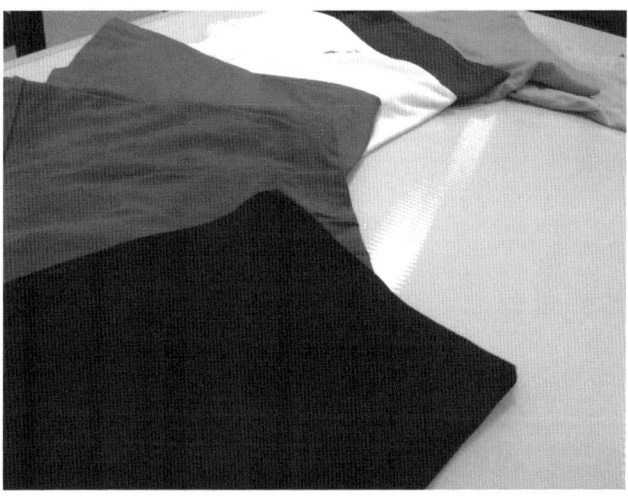

Figure 1.   Different coloured T-shirts.

Description of the Activities

Activity 1: Determining the temperature gain of the T-shirts

Procedure

1. Fix the position of the thermosensor above the heat source (Figure 2).
2. Connect the thermosensor to the multimeter.
3. Allow the heat source to achieve thermal equilibrium (stable temperature). Use a thermosensor to measure the temperature. The reading should only be recorded when it no longer fluctuates.
4. Then, place one layer of a T-shirt over the heat source (Figure 3) and start the stopwatch. Record the reading of the thermosensor at time, t = 0 min.
5. At every 1-minute interval, record the reading of the thermosensor in Table 1.
6. Repeat Steps 3–5 for subsequent measurements of different coloured T-shirts.
7. Plot the graph of sensor value against time for each coloured T-shirt (Figure 4).

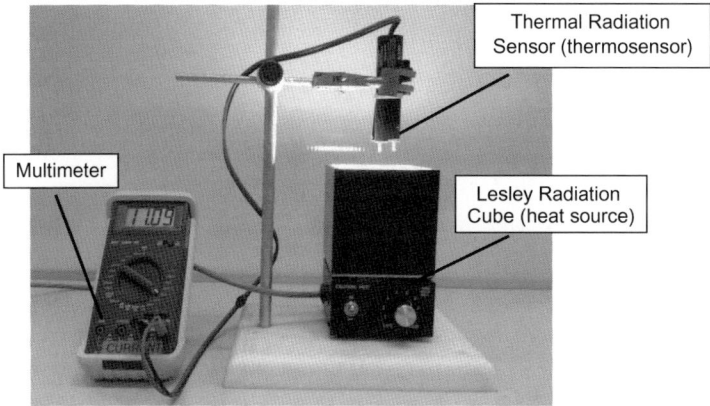

Figure 2.   The experimental setup for the multimeter, thermosensor, and heat source.

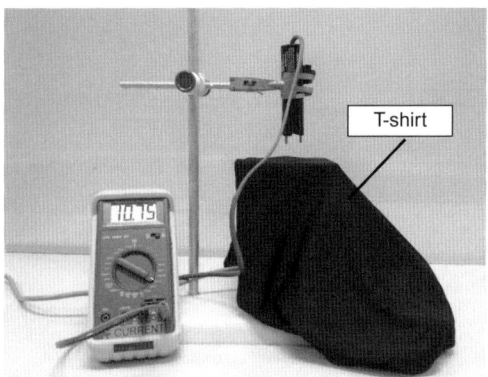

Figure 3.   The experimental setup for recording the thermosensor reading of the T-shirt placed over the heat source.

Assumption

The temperature of the surrounding is constant.

Results

Table 1.   A sample table showing the tabulation of experimental data at fixed time intervals for different types of T-shirt.

		Sensor readings						
		Initial reading [10 kΩ]	t = 0 min [10 kΩ]	t = 1 min [10 kΩ]	t = 2 min [10 kΩ]	t = 3 min [10 kΩ]	t = 4 min [10 kΩ]	t = 5 min [10 kΩ]
Material	**Colour**							
Cotton	White	17.76	12.60	13.53	13.76	13.83	13.93	13.97
Cotton	Orange	17.80	12.25	13.78	13.95	14.03	14.09	14.11
Cotton	Red	17.75	12.48	13.92	14.13	14.16	14.20	14.26
Cotton	Purple	17.38	12.31	13.80	13.91	13.97	14.04	14.04
Cotton	Blue-grey	17.75	12.31	13.82	14.01	14.08	14.14	14.11
Cotton	Light grey	17.76	12.27	13.77	13.90	13.93	14.03	14.03
Cotton	Black	17.77	12.54	14.23	14.30	14.40	14.43	14.43

Activity 2: Determining the darkness of the fabric

Procedure

1. Place a T-shirt in a controlled environment where it is well-lighted with minimal change in lighting condition.
2. Ensure that there are no shadows blocking the T-shirt.
3. Using the manual mode, set the camera to the following settings:
   a. ISO 200
   b. Aperture setting off 4.0
   c. Shutter speed of 1/250
   d. Lowest resolution, e.g., 640 × 480 px
4. Secure the camera on a tripod and fix its position.
5. Set the timer for the camera to capture a picture of the shirt.
6. Repeat Steps 4–5 for other T-shirts.
7. Download the free software, ImageJ from http://rsbweb.nih.gov/ij/download.html.

   ImageJ is a software that is used to analyse pixels in the image for the amount of light that is reflected, which can be done easily using the histogram function in the programme.
8. Open the image file (e.g., jpeg) of the shirt using ImageJ (Figure 5).
9. Use the rectangular tool to select an area of the photo where the colour is even (Figure 4).
10. Select from the menu bar: Analyse > Histogram. A histogram of the picture or the area selected would be produced (Figures 6 and 7).
11. Record the "max value" of the histogram. This value gives an indication of "darkness" of the colour, i.e., the smaller the number, the darker the colour and vice versa (Table 2).
12. Determine the order of shirts in terms of increasing *darkness*.

Extension

Inquiry activities

1. Explain if a thermometer can be used to record the temperature readings in the above experiment.

2. How can the rate of heat absorption and emission of different T-shirts be determined?

3. How would the temperature gain and darkness number differ for different types of materials? Based on the results, which type of material and colour would be most suitable for various climate?

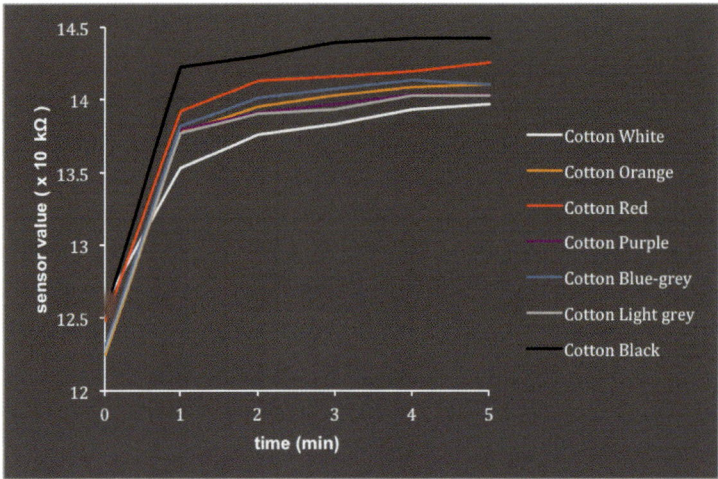

Figure 4.   A sample plot of sensor value against time based on real data. The data shows the increasing order of heat absorbed and emitted as follows: white, light grey, purple, orange, blue-grey, red, then black.

Figure 5.   A photograph of the T-shirt. The picture file was opened in ImageJ.

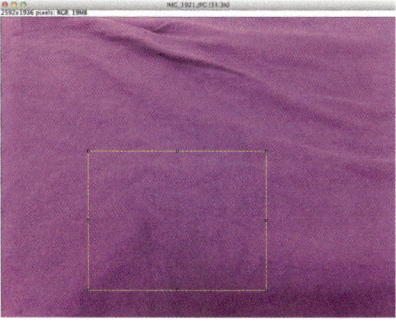

Figure 6.   The toolbar in ImageJ. Click on the rectangle option (circled). Click and drag over an evenly lit area on the photograph.

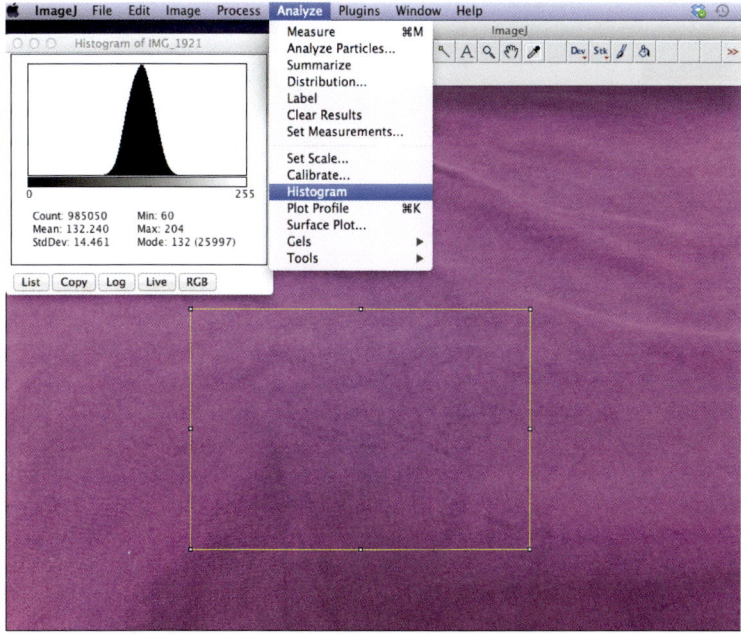

Figure 7.   Select "Histogram" under the "Analyze" menu for histogram analysis of the selected area in the photograph.

Table 2.   Results of the histogram analysis — darkness number of each coloured T-shirt. The results show that blue-grey is the darkest, followed by purple and then light grey. The results correspond with the experimental data in Activity 1.

Colour	Darkness number
Purple	219
Blue-grey	207
Light grey	254

References and Additional Recommended Resources

http://scienceline.ucsb.edu/getkey.php?key=1464
http://www.colormatters.com/color-and-heat-absorption

Notes for Users (if any) e.g., Hazards, Possible Problems

Students should be cautious when using the Lesley Radiation Cube to avoid any burn injuries.

Chapter 11

Teaching Kinematics Using Shuttle Run

Tang Chi Sin and Srinivasan Shyam

Introduction

In Singapore, the National Physical Fitness Award (NAPFA) Test (Fitness Assessment Singapore) is a compulsory physical fitness assessment of school students. NAPFA was first started in 1982 in conjunction with Singapore's *Sports for Life* initiative. Thereafter, it was adopted by the Ministry of Education and remains to this day, a key yardstick for the level of physical fitness of students. Under the new Physical Education syllabus, students will be assessed at Primary 4 and 6, Secondary 2, 4, or 5, and Junior College Year 2 or Pre-University Year 3 instead of annually.

The six components of the fitness test at primary level are sit-up, standing broad jump, sit-and-reach, shuttle run, inclined pull-up and 1.6 km walk-run. For male students aged 15 and above, they do vertical pull-up and 2.4 km walk-run. Apart from encouraging students to maintain a healthy and active lifestyle by providing a clear set of standards for them to work towards, the NAPFA test also serves to prepare male Singaporeans for the physical demands of National Service (NS) which they will embark upon completion of their secondary, tertiary, or pre-university education. The NAPFA Test has thus become a common thread that connects all Singaporean school students and features prominently in the range of common experiences of people in Singapore.

In this chapter, the use of an activity based on the 4×10 m shuttle run in the teaching of kinematics is explored. The shuttle run assesses an individual's speed, agility, and overall body coordination. The shuttle run is conducted in a 10 m stretch of track at the end of which are placed two small wooden blocks. Participants are required to dash from the start line to the 10-m mark, and pick up one block at a time and return them to the start line, covering a total distance of 40 m.

The shuttle run is ideally suited for use as an introduction to kinematics as it involves the following states of motion — rest, acceleration, constant speed, deceleration, and change of direction. This is the entire spectrum of motion of objects that students are required to identify and represent using graphs. Students' familiarity with the activity is expected to help them make connections between real-world motion and its representation using graphs.

A Closer Look at Kinematics

Kinematics is a subtopic in Newtonian Mechanics, which is the branch of Physics concerned with the analyses of the motion of objects. The central concepts in Kinematics are distance and displacement, speed and velocity, and acceleration.

A key component of kinematics taught at the secondary level is the graphical analysis and description of motion. Students are required to be able to plot and interpret distance-time, speed-time, and velocity–time graphs (SEAB, 2013). Distance– and speed–time graphs will be the focus of the activity in this chapter.

The activities below allow the learning of Kinematics through the use of NAPFA Test Shuttle Run. Not only is it something that Singapore students are familiar with, it provides a kinaesthetic representation for students to observe motion of a body at a constant or non-constant velocity. Thereafter, concrete representation can be translated into the form of distance– and speed–time graphs which in turn will provide a smoother transition to the learning of other concepts involving Kinematics. Through this activity, students are better able to have a better grasp and understanding of the concepts by navigating between these two main modes of representations.

The speed–time graph

There are a few features of speed–time graphs that students are expected to deal with:

1. Equate the instantaneous acceleration of an object to the gradient of the speed–time graph of its motion.
2. Relate the area under a speed–time graph with the distance travelled by the object. (*Note*: At the secondary level, students are only required to calculate distances for cases with uniform speed and uniform acceleration.)
3. Identify the rest, uniform motion and motion with uniform acceleration of an object based on the shape of the speed–time graph.

Details of the activity

- Objectives of the Activity
  The primary objective of the activity is to use the shuttle run as an example for describing motion using speed–time graphs. The activity involves capturing the motion of a person, as he/she completes one lap of the shuttle run, in a series of photographs. One lap includes dashing from the start line to the 10-m mark, turning around and returning to the start line. Further analyses can be conducted based on the captured images, and a speed–time graph of the motion can be plotted based on the data obtained.

- Materials Required

1. A 10-m stretch of track or flat surface
2. Measuring tape
3. Masking tape to make markings at 1-m intervals
4. A digital still camera with a time-lapse mode for capturing the person's motion at equal time intervals. Alternatively, a digital video camera to videotape the motion.
5. Image/Video Processing Software (e.g., Adobe Photoshop) for motion diagrams
6. Microsoft Excel

Figure 1.   10-m stretch of tracks with markings at 1-m intervals.

- Procedure

Part A: Gathering of Data

1. Form groups consisting of 4–5 members.
2. Each group will have a digital camera with a time lapse function.
3. Each group will record the motion of one team member using the digital camera. (*Note*: Care must be taken to ensure that the camera is kept stationary throughout the capture. This can be done using a tripod or by placing the camera on a table or any other flat, stationary surface.)
4. For the trial run, students will run at a steady speed from the start line to the 10-m mark, and the restriction will be lifted when they return to the start line from the 10-m mark.
5. Having collected the photographs, students will work on a laptop with animage processing software installed. This is so that a motion diagram can be created by merging the photographs together.

Part B: Data Analyses

1. Below are some examples of the processed motion diagrams. Note that the masking tape markings at 1-m intervals are visible in the processed images.

2. For the first part of the data analysis, tabulate the distance travelled in each interval of one second for the first of the motion diagrams as shown in Figure 2. Since the diagram shows uniform motion, the distances obtained should be equal.
3. Calculate the average speed for each interval. The value obtained should be identical. Here, average speed equals instantaneous speed.
4. Once this is done, a graph of speed against time can be plotted. The distance– and speed–time graphs are plotted as in the following diagrams:

- Motion at Non-Uniform Speed

Following this, students will consider the representation of non-constant motion through distance- and speed–time graphs. Students will repeat Steps 1 to 4 in the analysis of the person (shown in Figure 4) running at a non-uniform speed.

Figure 2.   Motion diagram for the motion of the runner from start to 10-m mark. Notice that the distance covered in each time interval is more or less the same (≈ 2 m).

Table 1.   Tabulation of distance travelled and the instantaneous speed by the runner at the end of each one second interval for the first 10 m of the shuttle run.

Time/s	Distance/m	Speed/m s^{-1}
0	0	2
1	2	2
2	4	2
3	6	2
4	8	2
5	10	2

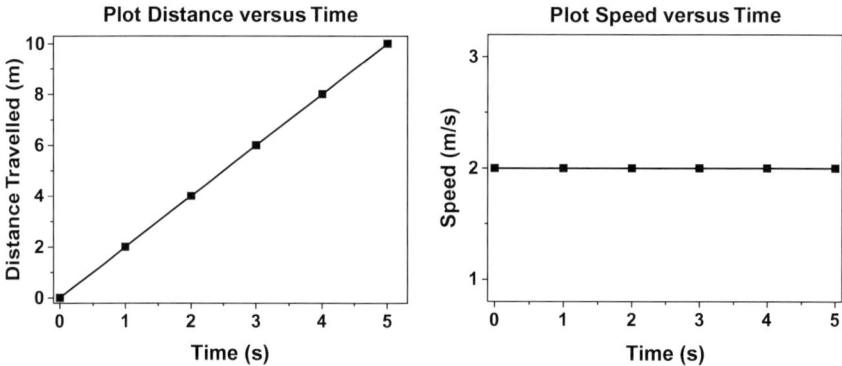

Figure 3.   Distance and the corresponding speed–time graphs depicting the motion of the runner at a constant velocity of 2 m s⁻¹.

Figure 4.   Motion diagram for the motion of the runner from the 10-m mark back to the start line depicting a non-constant motion. Runner dashes across the finishing line at the end of the run.

The motion of the runner at one second intervals is recorded in Table 2. Unlike the previous analysis, the instantaneous speed of the runner will not be recorded immediately. Instead, the distance–time graph is first plotted as shown in Figure 5.

What can be deduced about the speed of the person? How do you deduce this? The graph obtained is not a straight line as the speed is not constant. This is coherent with images of the runner's motion. In Figure 4, the picture on the right shows the speed–time graph for the second motion diagram.

There are a few points that need to be taken note of. First, the average speed and instantaneous speed are different. Second, the average speeds for the different time intervals can no longer be used as inputs for the speed–time graph, as the speed is constantly varying from one instant to

Table 2. Tabulation of the distance travelled by runner at the end of each one second interval for the returning 10 m of the shuttle run. Instantaneous speed is recorded in the third column after a series of discussion is performed with the students.

Time/s	Distance/m	Speed/m s⁻¹
0	0	0.5
1	1	1.5
2	3	2.5
3	6	3.5
4	10	4.5

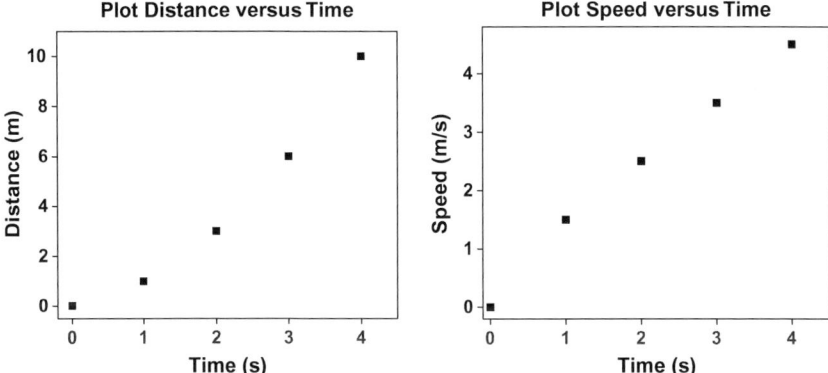

Figure 5. Distance and speed–time graphs of runner for the returning 10-m of the run. The distance travelled in each time interval is increasing. It can de deduced that the increasing speed resulted in the greater distance completed within the same amount of time.

another. This is unlike the first case where it remained constant throughout.

What does the gradient of the speed–time graph tell you? How do you know?

The gradient of the speed–time graph equals acceleration. The straight line graph obtained is indicative of constant acceleration.

What does the area below the graph tell you? The area under the graph indicates the distance travelled by the object.

Calculate the distance travelled at the end of each time interval and relate this to the motion diagram.

- Extension Inquiry Activity — Vector quantities of displacement, velocity, and velocity–time graph.

Thus far, the change in direction has not been treated very carefully as only the scalar quantities of distance and speed have been discussed using the activity. Use the two phases of motion shown in the two motion diagrams and represent both phases using a velocity–time graph.

The essential difference is that the entire graph will lie below the time axis of the velocity–time graph since motion in the second phase is oppositely directed to motion in the first phase.

References

Fitness Assessment Singapore (2003). National Physical Fitness Award (NAPFA). Retrieved on June 12, 2014 from http://fitnessassessmentsingapore.com/napfa-test/.

Singapore Examinations and Assessment Board (SEAB) (2013). 5058 Physics (with SPA) Ordinary Level 2013. Singapore.

Authors' Reflections

BOO Michelle

I hope to encourage my students to recognise the range of applicability of science in the real world. Science is not limited to the laboratory lessons in school or sophisticated laboratories in a research facility. Neither are scientific experiments restricted to the use of apparatus and chemicals found only in the laboratory. Most scientific knowledge originates from simple ideas and after careful experimentation and observations, concepts are then formally developed. By applying science concepts to explain everyday phenomena, one will realise that science is not an abstract, complex topic but is, in fact, a fun and exciting one! I earnestly encourage everyone to be more aware of science in their daily life. I wish all readers a *fruitful* journey in the lifelong appreciation of science.

CHEW Shuhui Eunice

Science is all around us but we often fail to relate the teaching of science to our culture. When doing school experiments, students would often ask questions such as, "How is this experiment related to us?" and "Why do we need to conduct this experiment?" Students are searching for meaning in what they do. As educators, we need to find ways to better engage these students by making science meaningful and fun. As I was designing culturally relevant activities for the book, I discovered that there were many ways to relate science to our culture. In order to do so, we need to make a conscientious effort to discover and explore. This is possible by asking questions, such as "Does the pearls in my bubble tea contain starch?" and "Is it possible to separate milk from Teh-C?" What began as a question bore an idea which developed as a hypothesis. This writing experience had reshaped my understanding of teaching science, one that is no longer restricted to standard school experiments. Rather, it extended to experiences that relate to people living in Singapore. Science became more personal. This experience had challenged me as an educator to conscientiously create meaning in science learning by specifically relating it to our culture. It is my hope that students can better understand and appreciate the things around them and what they do as they learn culturally relevant science.

CHO Wen Jing

I found the process of writing a lesson activity that was relevant and connected to students' cultural context quite challenging. In trying to come up with ideas, I had to consider the daily experiences common to Singaporean students and identify how Science concepts can be applied in these experiences. Typically in the teaching of concepts, teachers tend to focus on the learning outcomes and then make use of whatever contexts that are applicable and helpful for students in their learning. However, my co-authors and I approached it the other way by foregrounding culture and I think it taught me to see science in things rather than apply science to things. I was personally challenged to "see" the science in daily phenomena. Perhaps students can be encouraged to think about their own experiences and everyday observations with a curious and inquiring mind. This would in turn motivate and interest them in the subject of science and its various disciplines, and henceforth, to discern and learn through their own questioning mind.

GAN Ghim Kui

The journey in searching for a culturally relevant science activity in Asia is a fun and fruitful prospect. The purpose of my work in this book is twofold: to create an effective teaching and learning activity, as well as to create a relational purpose to real applications characteristic of the Asian culture. The relational element creates a sense of familiarity to students that would be proximally more constructive in their learning. At the start, it was challenging to come up with an activity that could be taught in the field of elementary physics. The initial path was fraught with disappointments due to over-complexities in an activity we initially planned, which was the radio-making experiment. As such, we strove to think of a more simple, yet elegant experiment. The availability of materials and skill sets required to conduct an experiment are critical factors for the activity design. For example, enamel wire is easily available in the United States but not in Singapore. This led us to the design of an activity that is very applicable in the Singapore context and other tropical countries in Southeast Asia. The experiment employs the use of materials including T-shirts, data-loggers and software. I felt that the use of data loggers and software is a highlight in this activity as it allows students to use technology in scientific investigations, which is highly in tune with the current age and paradigm.

KHOH Rong Lun

Even though I have only completed less than two years of formal classroom teaching, I observed that there are two common requests from students when I asked them how their learning experience can be improved: (1) help students see the relevance of the knowledge they learn, and (2) have more hands-on-work. It is precisely because of these feedbacks that motivated me to be on board this journey where I have a taster of writing curriculum that will pique students' interests for science and to help them develop greater love for the discipline. Writing science activities anchored in cultural relevance is by no means an easy task as it calls for a deep understanding of one's culture. In writing this activity, I learned that the same science concept can be told through the perspective of a generic textbook written with an evident "western-centricity" or shared through a distinctly Singaporean lens which would probably resonate more strongly with local students. With a higher level of cultural relevance, there is potentially greater student engagement and enrichment of their learning journey. I would like to thank Dr. Teo Tang Wee for inviting me on board for this special project as I see myself growing personally and professionally through this experience.

KOH Bing Qin

Science is a subject which many students find intangible, but ironically, science is all around us. In fact, we are applying scientific knowledge subconsciously in our daily life. Prior to the writing of this book chapter, I had always relied on the use of common reagents and solvents to illustrate concepts in class. In the process of completing this work, however, I realised that day-to-day materials can possibly be as effective in bringing out the essence of simple experimental techniques. It is not necessary for us to use sophisticated equipment or perform impressive chemical reactions all the time as simple items such as flowers and pastries can probably arouse interest among students too. In the course of writing the book chapter, I recognised that infusing culturally relevant elements in teaching and learning provides a very good avenue for inquiry learning and discussions beyond the syllabus content. While reading on the ingredients of Peranakan pastries, I was drawn to the term 'anthocyanins' and noted the possibility of using butterfly pea flower extract as a pH indicator. In the same way, I believe that students will be enticed to read up more as they notice that items which they can relate to are being applied in the school laboratories.

LIM Jia Ying Jessica

When my co-authors and I were conceptualising this chapter, our main concern was to challenge students to think more critically about objects or phenomena that were familiar to them in their daily lives. The average consumer often consumes without giving much thought to the product being used. An example of this is the use of EZ-link cards that are used to pay for taking public transport simply by tapping the card on a reader. It is easy to forget that such a simple tool makes use of physics and engineering knowledge. This writing experience had helped me to be more aware about the application of science around me and identify ways to include these examples into my science teaching. One particular challenge that we faced was in identifying a science-related idea that is uniquely local to Singapore as most examples we found were more relevant to the western context or required complex instrumentation and sophisticated science understanding. For example, our first attempt was to build a radio using materials suggested from the web. However, we quickly realised that most of the materials needed could not be found in local stores, hence making the project unfeasible to local students. Although this is the first time I have encountered a project related to culturally relevant science activities, I believe that this idea will continue to flourish as students today learn to take ownership of their own learning.

LIM Shan Yan

As a fledging teacher with fresh ideas and enthusiasm, I leapt at the opportunity to contribute a chapter to this book. It is truly an excellent platform for teachers to explore various ideas and make the conscious effort to teach science in culturally relevant ways. While I acknowledge that some students may not have a strong passion for science, teachers should strive to look for innovative ways to create memorable learning experiences for students. My initial inspiration for the book chapter did not come from the coconuts. My co-author and I wanted to explore the idea of investigating the anti-inflammatory effects of saga seeds. However, the experiment could not be carried out in a school laboratory. We continued to look for new ideas and designed one based on coconut water. Writing culturally relevant science activities had reshaped my understanding of science teaching and learning. It had made rethink the possibility of teaching science in a less deliberate manner by capitalising on the students' own experiences.

LIN Jiansheng

Having been involved in this curriculum writing has prompted me to recall my experience of culturally relevant science as a student. While there may seem to be no recollections of such reference made in my lessons it was, in fact, carried out unknowingly in the classroom by my teachers. For example, science teachers would share the use of pandan leaves by taxi drivers to ward off pests such as cockroaches while introducing the usage of volatile organic compounds in aromatic and perfumes industries. Considering the growing need to engage our 21st century learners, culturally relevant science becomes a useful strategy in making their learning experience more authentic and meaningful as it provides students with opportunities to connect their learning experiences to their everyday lives. This would require educators to make a conscious effort in weaving culturally relevant science into their teaching of science as a body of knowledge by making references to students' diverse cultures, backgrounds and personal experiences.

LOW Wei Chuan Matthias

It had been an enriching experience to be involved in this curriculum writing project on culturally relevant science as it had made me more aware of the importance of bringing in local examples that students can resonate with. As I reflected on my own teaching experiences, there were instances where examples presented in a lesson did not resonate with my students due to their backgrounds. Teaching is also about knowing the needs of students and tailoring the lessons to meet their learning needs. I wondered: How then can I bridge the gap between my lessons and students' experiences? After doing some research and reading for this project, I realised that one way to connect students to the scientific concepts taught in the classroom is to introduce cultural elements that students are familiar with. They will form connections between daily occurrences and abstract concepts that they learn in the classroom. When they are able to forge such connections, the retention of knowledge could be strengthened. At the same time, it will help heighten their interest in the learning of science as they are now able to explain phenomena that they encounter with what they learn in the science classroom.

NG Shi Han

I used to believe that science is a part of our daily lives — in the interpretation of weather conditions, invention of technological gadgets, improvement in healthcare etc. However, I did not know that science has anything to do with cultures. However, after considering how the local cultures can be brought out using science experiment, my perception shifted. Following this experience, I often asked myself how I can further arouse the curiosity of my students to learn science and, at the same time, develop better understanding of the local culture. This inspired me to think about using inquiry-based teaching approaches in my classroom. I will look forward to sharing this experience with my colleagues and students.

SRINIVASAN Shyam

This was the first time I am involved in writing a book chapter that is published. Naturally it filled me with excitement and I was glad to be able to work on this with my friend Chi Sin. We came up with our plan for the chapter based on one of our NIE assignments on kinematics. Our initial ideas were based upon motion diagrams used in textbooks. However, we decided to contextualise the activity using the shuttle run component of the NAPFA test. Also, motion diagrams are typically not done by students, so we thought it may be good for them to make motion diagrams. In the process they can make connections between concrete and abstract ideas. Overall, I would say it was a rewarding experience and I built on this thought exercise and used the activity to teach kinematics in my school as well. It was successful in some ways, but I hope to improve on it and use it in subsequent years. My sincere thanks go to Dr. Teo and Mr. Khoh for their understanding and flexibility with deadlines, as well as valuable comments and feedback. Most of all, I would like to thank them for giving us this opportunity to contribute to their book, a real privilege indeed, especially as a fledgling teacher.

TAN Yong Leng Kelvin

Looking back at the footprints in the sand…

When Tang Wee asked me to contribute a chapter for this book, I was thrilled by the challenge. Although I am an avid reader of science books written for the public, my previous experience consisted solely of writing the manuscripts of my chemistry research work for publication. Would I be able to follow in the footsteps of my favourite authors, by communicating my ideas simply and effectively to the target audience of secondary school science teachers and students?

Over the next few weeks, I contemplated several themes: What would be appealing yet stimulating for secondary school students? Tang Wee suggested exploring biodegradable materials; I was intrigued and began to examine this topic in greater depth. Hwa Chong Institution (HCI), my alma mater and the school that I am currently teaching at, has been embarking on a Green Plan for several years now. One of the earliest initiatives implemented was the use of biodegradable tableware for all catered meals at school functions. Were the forks, spoons, plates and cups that my colleagues, students and myself have been using really *biodegradable* and *environmentally friendly*, as their manufacturers so claimed? Throughout this journey, I had the opportunity to engage in many interesting and insightful conversations with my colleagues and students at HCI, as well as the people I met in Switzerland while I was on sabbatical. I am very grateful to all of them for sharing with me their knowledge and understanding of environmental issues. Colleagues from the HCI Science Department helped with giving suggestions and clarifying my doubts on Biology and Physics concepts; I also had the privilege to work closely with the science laboratory technicians to trial the experiments described in this chapter. I hope that our combined efforts will inspire students of science, both young and old, to work together and seek novel solutions for current and future challenges.

This chapter is dedicated to Mrs. Tham-Kee Yong Huang, an outstanding chemistry educator, mentor, and colleague, who has been a leading light for Hwa Chong's green initiative since it began. Mrs. Tham, may you continue to be an inspiration to your colleagues and students, as well as a champion for the green movement!

TANG Chi Sin

It is generally agreed that students tend to encounter difficulty in relating speed–time and distance–time graphs from the graphical and mathematical point of view. By incorporating a real life and more visual component into the study of Kinematics, it may better facilitate students' learning by engaging what they have encountered in real life to what is to be taught in the textbook. In getting students involved in the production of motion diagrams, there is also an added element of fun to their learning experience. The intent of this project is also to inject some form of novelty to the learning process. The aspect of learning by doing should better engage students in the learning of Physics. I would like to thank my co-author, Shyam Srinivasan for roping me in to contribute to this book chapter. It has been a fulfilling and fruitful learning experience in painstakingly thinking through and selecting the various modes of Physics instruction in the topic on Kinematics. Special thanks also to Dr. Teo Tang Wee and

Rong Lun for their invaluable advice in revising the learning activity so as to create a more fulfilling learning experience for students.

TEO Tang Wee

This is an inaugural attempt to engage Singapore teachers in curriculum writing of culturally relevant science activities for students. Initially, I was not confident that I would get enough chapters to fill the book because it required a change in mindset — teachers are curriculum writers and science as culturally relevant. It is my strong belief that culture must be foregrounded in attempts to connect science to students' everyday lives. Since the idea of culturally relevant science is new to most, if not all, Singapore teachers who themselves had not experienced science learning in this way, they had to relinquish their prior learning experience to rethink how science could be learned differently. I was heartened that the response to the call for book chapters was spontaneous and many of my current and former colleagues including pre-service and in-service teachers took up the curriculum writing task. Teachers can adapt the ideas in this book in their own science classrooms to help students apply science concepts to their everyday lives and not simply acquire scientific knowledge as canons of knowledge. Teaching science in culturally relevant ways, I argue, is enroute to achieving scientific literacy.

Subject Index